W9-BNO-754

LEGO® MINDSTORMS® NXT
ONE-KIT WONDERS

LEGO® MINDSTORMS® NXT ONE-KIT WONDERS

ten inventions to spark your imagination

james floyd **kelly**, matthias paul **scholz**, christopher r. **smith**, martijn **boogaarts**, jonathan **daudelin**, eric d. **burdo**, laurens **valk**, **bluetoothkiwi**, *and* fay **rhodes**, *editor*

no starch press

12 11 10 09 08   1 2 3 4 5 6 7 8 9

ISBN-10: 1-59327-188-3
ISBN-13: 978-1-59327-188-6

Publisher: William Pollock
Production Editor: Megan Dunchak
Cover and Interior Design: Octopod Studios
Copyeditor: Jeanne Hansen
Compositor: Riley Hoffman
Proofreader: Kathleen Mish

For information on book distributors or translations, please contact No Starch Press, Inc. directly:

No Starch Press, Inc.
555 De Haro Street, Suite 250, San Francisco, CA 94107
phone: 415.863.9900; fax: 415.863.9950; info@nostarch.com; www.nostarch.com

Library of Congress Cataloging-in-Publication Data

```
LEGO MINDSTORMS NXT one-kit wonders : ten inventions to spark your imagination / James Floyd Kelly ... [et al.].
      p. cm.
  ISBN-13: 978-1-59327-188-6
  ISBN-10: 1-59327-188-3
  1.  Robotics--Popular works. 2.  LEGO toys. 3.  Robots--Programming. 4.  Robots--Design and construction.
I. Kelly, James Floyd.
  TJ211.15.L44 2008
  629.8'92--dc22
```

                                                    2008041218

about the authors

BlueToothKiwi

BlueToothKiwi lives in New Zealand with his partner and four children. He has an electrical engineering background but is now working in the banking industry. He has been building robots with MINDSTORMS® NXT for two years and especially enjoys building outdoor robots that go places too dangerous for humans. BlueToothKiwi is a contributor to The NXT STEP blog and Brick Journal. He is also a member of the MINDSTORMS Community Partners (MCP), a group that assists LEGO® with testing and growing the NXT product. He also teaches local school children how to build and program NXT robots and mentors a Robocup team. He would like to thank Christine, Robert, and Daniel for all their help in the development of the Candy Picker (Chapter 1) and The Bike (Chapter 10).

Martijn Boogaarts

Martijn Boogaarts, a resident of Duiven, The Netherlands, is a freelance integration technology trainer. In 1986, he started a LEGO "robotica" club at his school, and he has since built many robots. Martijn was one of the initial organizers of LEGO WORLD. He has built several large demonstration models, including the Road-Plate-Laying Machine, a working car factory (using 27 RCXs), and a pinball machine. In April 2005, he contributed to the AFOL MINDSTORMS tournament in Billund, Denmark, and, later that year he was asked to join the MINDSTORMS Community Partners (MCP). Martijn contributes to The NXT STEP blog and shares knowledge about the NXT to show that you can build it, too. Martijn was a contributor to *The LEGO MINDSTORMS NXT Idea Book* (No Starch Press, 2007) and is the designer of BobBot (Chapter 5).

Eric D. Burdo

Eric Burdo is a grown-up (okay, a big kid) working as a computer programmer in Maine. He's been infatuated with robotics and creating with LEGO bricks since he was a little kid. When the RCX kits were produced, he purchased two. His wife bought him an NXT kit as a birthday gift in September of 2006, and he became a contributor to The NXT STEP blog a few months later. Eric likes to tinker with hobby electronics and teach his six-year-old how to dissect old electronic toys. He also teaches computers and LEGO robotics part time. He would like to thank Adrianne, Jacob, and the rest of the students of his Robotics and Engineering class for helping with the testing and building instructions for his robot contribution, The Hand (Chapter 7).

Jonathan Daudelin

Jonathan Daudelin is a 16-year-old robotics enthusiast from New Jersey. He co-authored *FIRST LEGO League: The Unofficial Guide* (No Starch Press, 2008), *NXT Robotics Competition Workbook* (self-published, 2007), and the popular *The LEGO MINDSTORMS NXT Idea Book* (No Starch Press, 2007). Jonathan helped start and was a member of a FIRST LEGO League team, Built On The Rock. In their second year of competing, he and his team won first place Robot Performance and first place Innovative Robot at the World Festival in the 2006–2007 Nano Quest challenge. His team's robot achieved perfect scores on all three of its competition rounds—something that had only been done once before in World Festival history. Jonathan created RoboLock (Chapter 6).

Jim Kelly

Jim Kelly is a freelance writer based in Atlanta, Georgia, with degrees in English and Industrial Engineering. Jim was accepted into the MINDSTORMS Developer Program (MDP) in 2006 and helped to beta test the LEGO MINDSTORMS NXT kit and software. He is currently a member of the MINDSTORMS Community Partners (MCP). Jim co-wrote *FIRST LEGO League: The Unofficial Guide* (No Starch Press, 2008) with Jonathan Daudelin and is the author of *LEGO MINDSTORMS NXT: The Mayan Adventure* (Apress, 2006) and *LEGO MINDSTORMS NXT-G Programming Guide* (Apress, 2007). Jim founded The NXT STEP blog in January 2006 and is the contributor of PunchBot (Chapter 2).

Fay Rhodes

For this project, Fay has taken on the role of cat herder—otherwise known as manager and editor. With the help of the staff at No Starch, she has taken pains to translate the excellent instructions from our authors into language comprehendible to the average American 11- to 13-year-old (and his or her grandparents). Fay was a contributor to *The LEGO MINDSTORMS NXT Idea Book* (No Starch Press, 2007) and is the author of *The LEGO MINDSTORMS NXT Zoo* (No Starch Press, 2008). She is also a member of the MINDSTORMS Community Partners (MCP) and is a strong advocate for using the MINDSTORMS NXT as an integrative teaching tool in schools. She is a recent immigrant to Perry, Oklahoma (from Massachusetts), where she is mentoring public school students on two new FIRST LEGO League teams. She would like to thank her husband, Rick, for discovering and introducing her to the MINDSTORMS NXT.

Matthias Paul Scholz

Matthias Paul Scholz is a mathematician, living in the Black Forest (Germany). Presently, he is a member of the LEGO MINDSTORMS Community Partners (MCP) and an official NXTpert. He is also the administrator and editor of the German NXT blog Die NXTe Ebene, one of the contributors to The NXT STEP blog, author of *Advanced NXT: The Da Vinci Inventions Book* (Apress, 2007), and co-author of *The LEGO MINDSTORMS NXT Idea Book* (No Starch Press, 2007). Matthias designed M, the M&M sorter (Chapter 3). He would like to thank his wife for her encouragement; Fay Rhodes for her patience with the team; Laurens Valk for his commitment in rendering and reviewing the building instructions for M; and all of the co-authors for their input and inspiration.

Christopher R. Smith

Christopher R. Smith (aka Littlehorn) is a senior quality assurance inspector in the Shuttle Avionics Integration Laboratory (SAIL) at NASA's Johnson Space Center in Houston, Texas. He invented an inspection tool recognized by NASA and has been recognized with the prestigious Space Act Award. He has been designing LEGO MINDSTORMS robots since rediscovering LEGO robots in 1997. Chris is one of the pioneer volunteer moderators of the MINDSTORMS website community forums, providing a safe online community for everyone who participates. Chris is a member of the MINDSTORMS Community Partners (MCP), a contributor to The NXT STEP blog and a contributor to *The LEGO MINDSTORMS NXT Idea Book* (No Starch Press, 2007). Chris is the creator of the NXT Dragster (Chapter 4). He would like to thank his wife, Veena, and his children, Revi and Benjamin, for their support and inspiration.

Laurens Valk

Laurens Valk is a 16-year-old resident of The Netherlands. He got his first MINDSTORMS kit in 2005 and has been a robotics enthusiast ever since. Two months after he got his first MINDSTORMS NXT set in 2007, he was invited to become a contributor to The NXT STEP blog. Laurens enjoys creating computer-aided design drawings of his creations and is responsible for the creation of the clear images in this book. He would like to thank Philippe Hurbain and Jaco van der Molen for helping him with some of the problems that occurred during the building image generation process. Laurens is the designer of the SPC (Chapter 8) and GrabBot (Chapter 9). Look for more of his designs on the Internet.

brief contents

contents in detail

acknowledgments

The authors would like to thank Kevin Clague, Travis Cobbs, Philippe Hurbain, and the LDraw community for developing the numerous graphic design and CAD tools that made this—and many other NXT books—possible. They would also like to thank Fay Rhodes, for her role as editor and manager. Thanks to her support and infinite patience, the authors were able to work together to create a book with ten unique and amazing chapters.

The editor would like to thank the contributors for their patience in a very long effort to bring you the best models we could come up with for the basic NXT kit. Working with eight different designers on three different continents is a challenge for everyone. I would especially like to recognize . . .

The folks at No Starch Press, for their willingness to take on another group project.

James Floyd Kelly, for his role as back-up manager. His advice and support has been invaluable in this process.

Laurens Valk, for his enthusiasm, tireless attention to detail, and the huge amount of time he spent creating the images for the building instructions in this book.

Laurens Valk, Martijn Boogaarts, Matthias Paul Scholz, and BlueToothKiwi, for the forbearance they showed when their work was reinterpreted for the American adolescent.

Chris Smith, for the construction of the dragster models on the book's cover.

BlueToothKiwi, for putting in a tremendous amount of extra hours to make it possible to build the NXT Bike with a single NXT kit. (His original design required the large motorcycle tires found in the Education Resource set.)

Eric D. Burdo and Jonathan Daudelin, for their immediate response, without complaint, to every single request.

CandyPicker: a candy-picking robot
with built-in generator and remote control

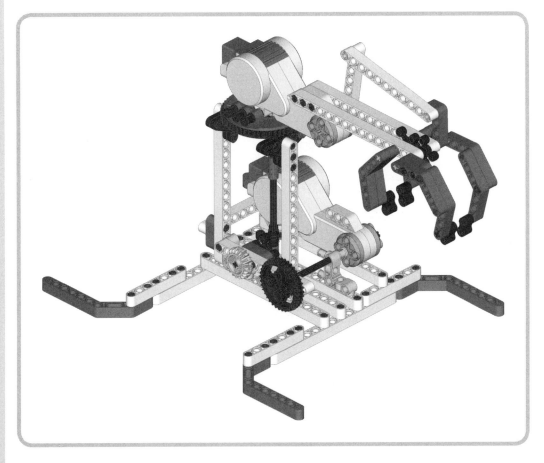

Figure 1-1: The CandyPicker

Remote-controlled robot arms have been in use for a long time. They have many practical uses, such as handling hazardous substances and industrial and medical waste. I designed this robot to pick up candy, but you can use your imagination to find other uses for it.

You may be surprised to learn that the CandyPicker generates its own power and does not require batteries, the NXT Brick, a computer, or programming of any kind. With these instructions—and your imagination—you will build a robot that can pick up all kinds of things, and all you have to do is turn two knobs!

When you have a design you like, invite some of your friends over and challenge them to a candy-picking contest. All you'll need to do is place a pile of your favorite candy on the floor or on a table, add a timer, and see who can grab the most candy in 30 seconds.

The arm with a motor-controlled gripper on its tip is controlled by a second motor, which acts as a generator. A knob in the base turns the second motor, which in turn controls the fine movements of the gripper on the end of the arm. The two motors are connected by the longest NXT cable.

By the time you have finished building the model, you should have a good understanding of:

Levers   How to turn circular movements into linear movement and do something useful with it

Gears   How to step down (reduce) and increase the power and how to transfer the motion from one plane (horizontal) to another plane (vertical)

building the CandyPicker

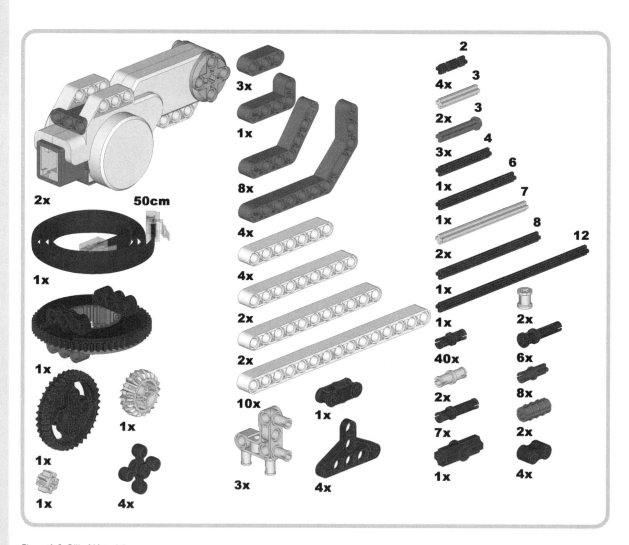

Figure 1-2: Bill of Materials

the base

1

6x

2x

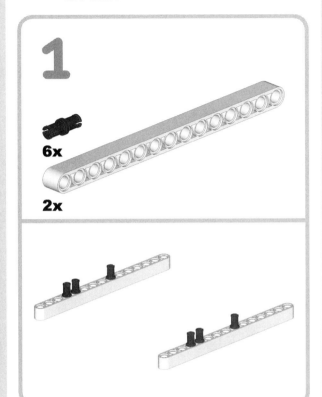

2

2x

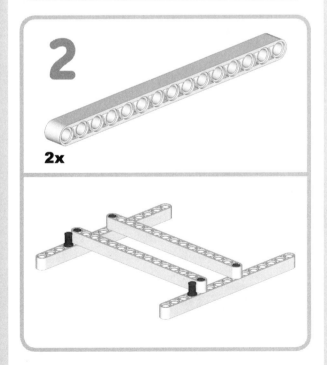

3

2x 3

1x

1x 1x

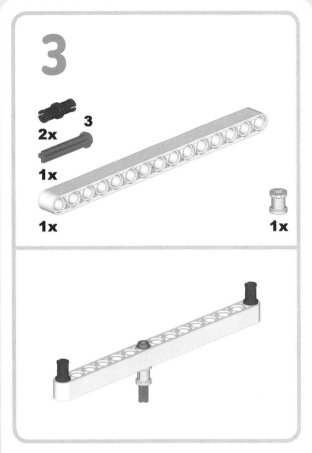

4

1x

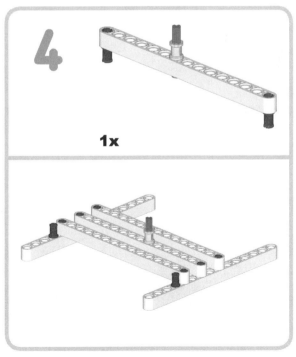

5

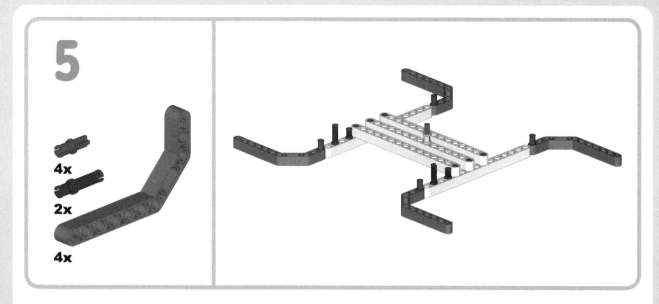

4x
2x
4x

Use blue pegs in step 5.

6

12x

7

4x

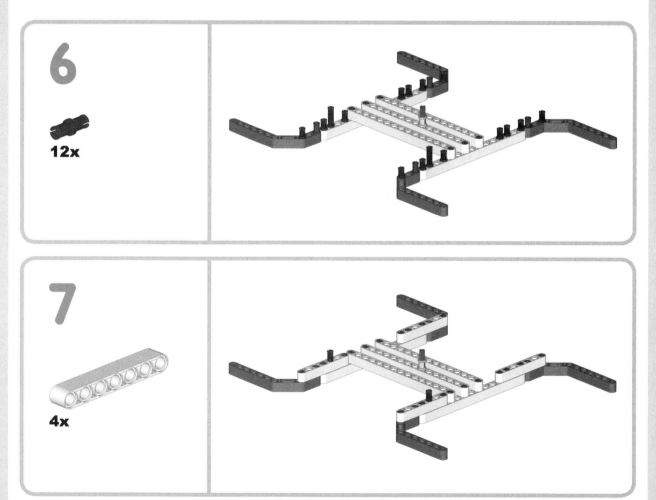

8

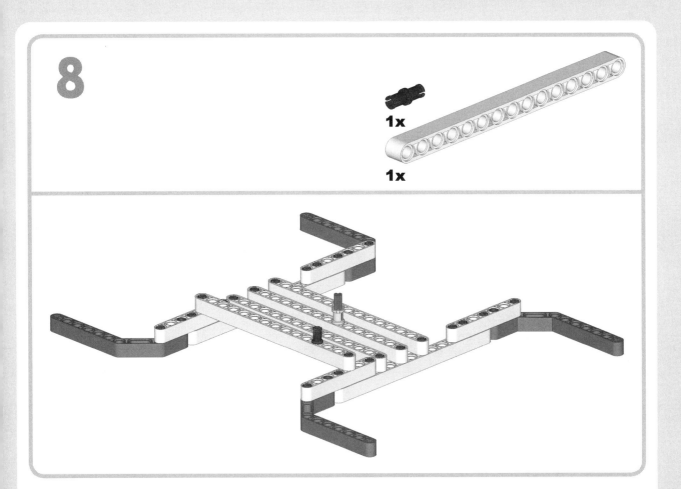

Now that you have completed the base, note how it spreads over a large area so that the tall tower you are about to build won't topple over as it lifts things. To make the CandyPicker even more stable, we'll place the first NXT motor on the base to help lower the center of gravity.

the tall tower and the turntable

1

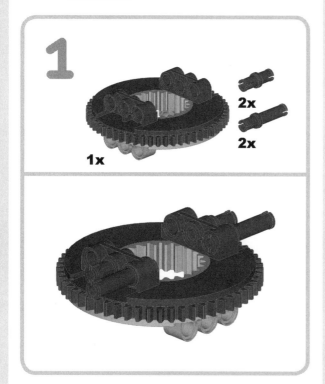

1x   2x   2x

2

4x

3

4x

4

2x

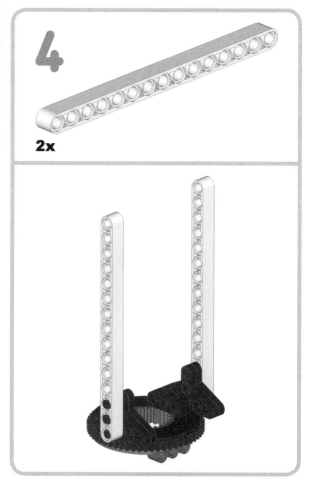

5

2x
2x

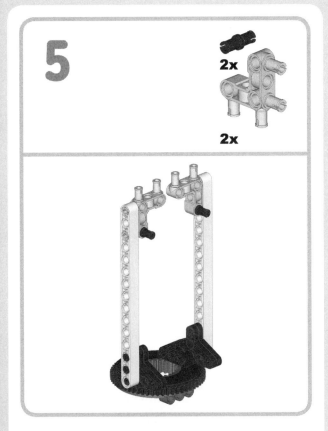

7

1x   4
1x

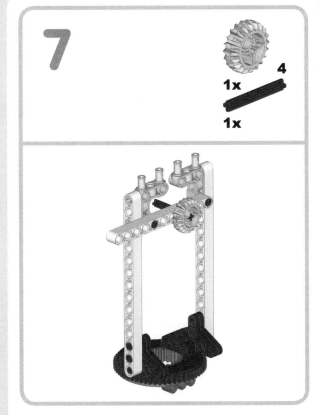

6

1x

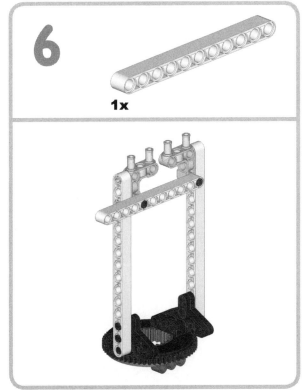

8

4x

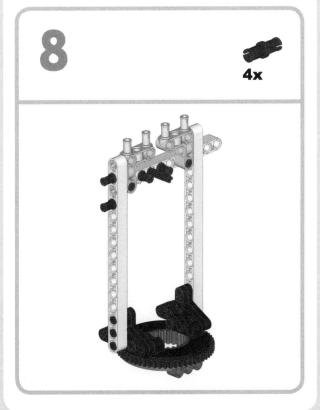

9

2x

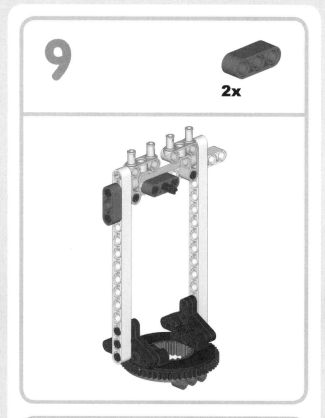

10

1x   1x

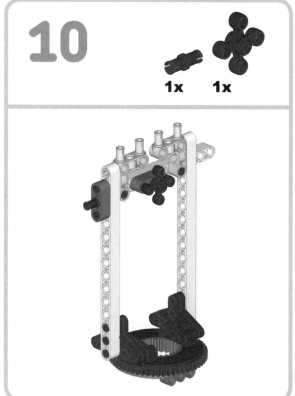

11

3

2x

1x

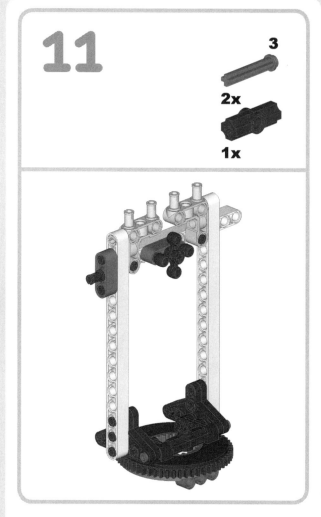

Now that you've built the base and the tower, attach the tower to the base as shown in the following figure.

12

1x

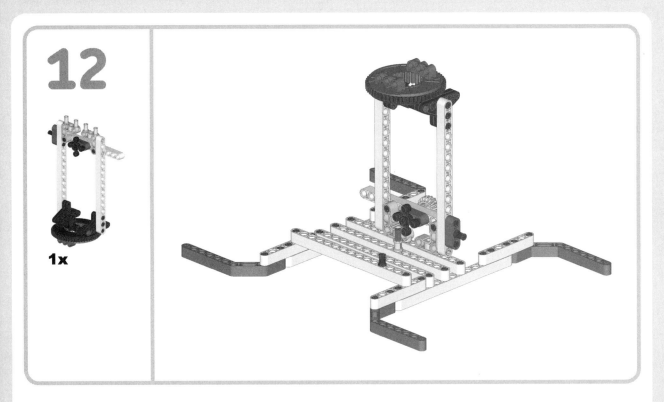

Now we use the 90-degree knob wheel to convert the horizontal motion from the base's controller to vertical motion, which is then transferred to the top of the tower.

Create a vertical axle as shown in steps 13 and 14, and add it to the base of the tower you have created, as shown in step 15.

13

8
1x
1x   1x

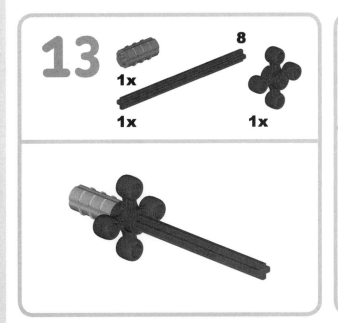

14

6
1x
1x

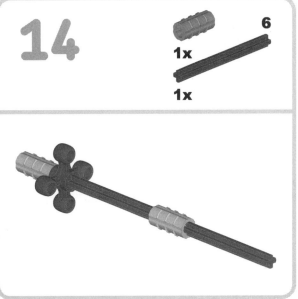

15

1x

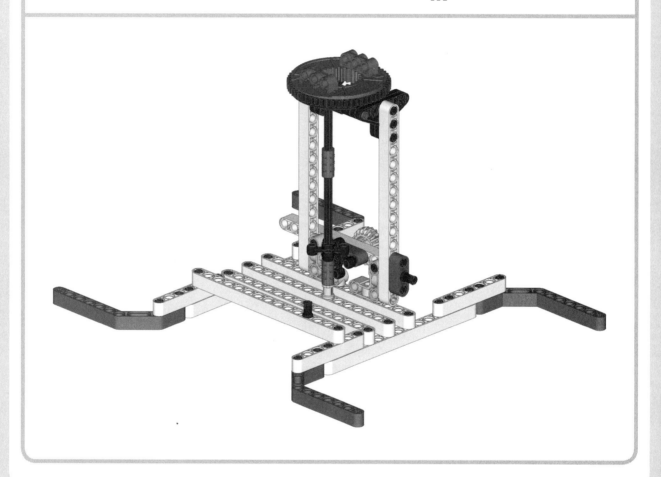

the generator

The NXT motor on the base behaves as a generator (or a dynamo); turning it makes it generate the electricity to drive the second motor that powers the gripper on top of the tower.

Assemble the motor according to the following instructions; attach it to the base, and add the gears and beams as shown.

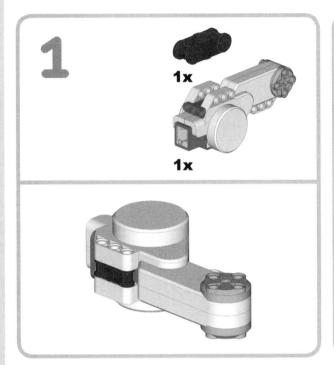

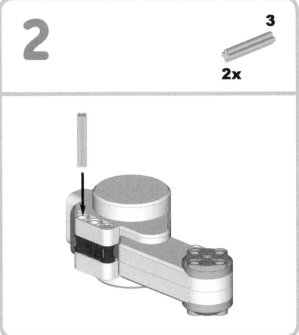

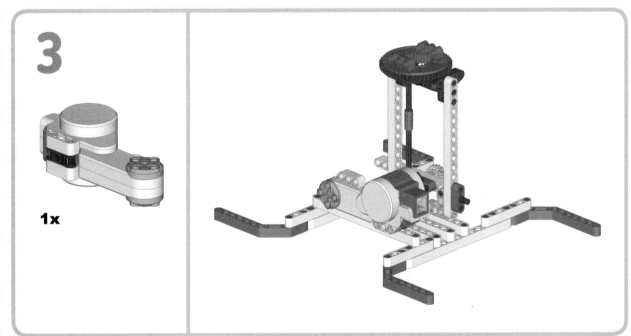

4

2x

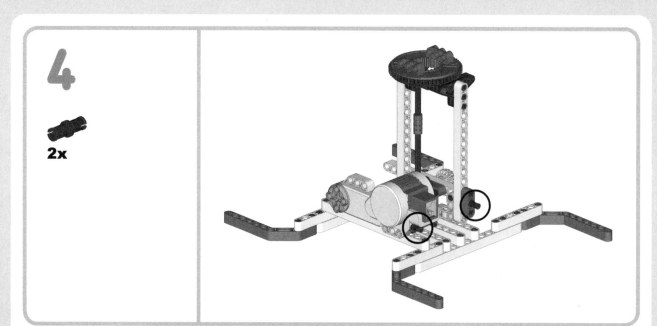

5

1x

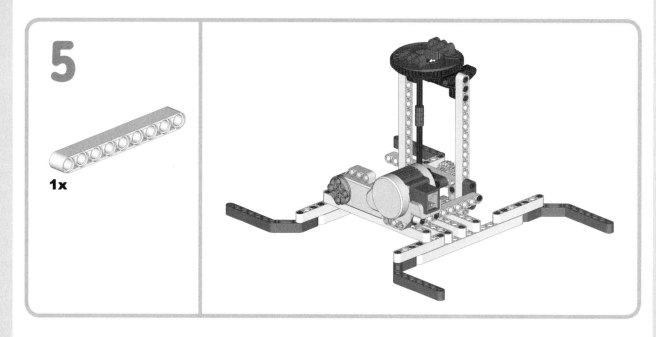

6

1x

1x

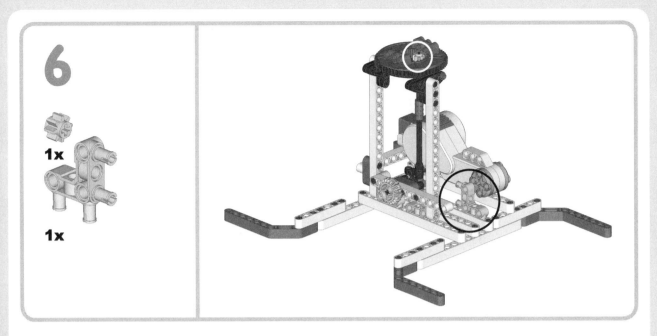

7

1x

12

1x

1x

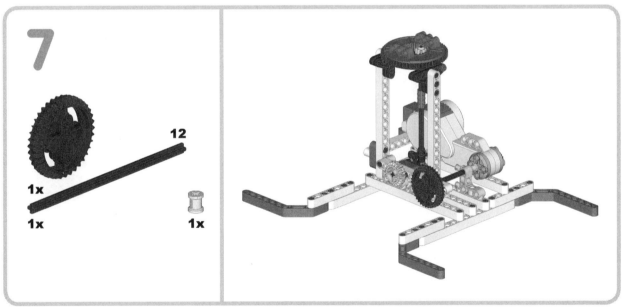

Having completed the base, you are ready to make the arm. The arm is a simple lever that translates the circular motion of the motor into linear movement, which in turn controls the gripping movement of the claws.

the arm

1

2x

7

1x

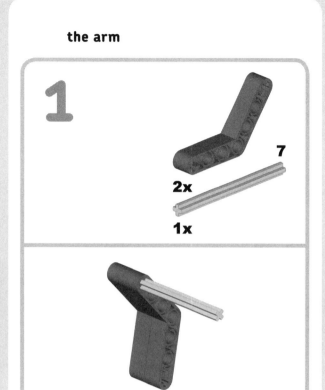

2

1x

1x

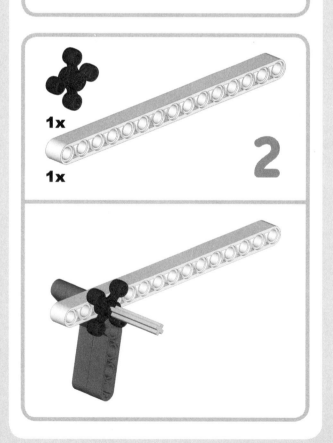

3

1x

7

1x

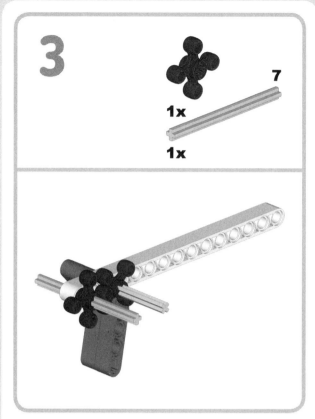

4

2x

1x

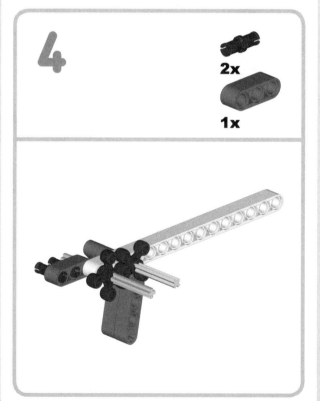

5

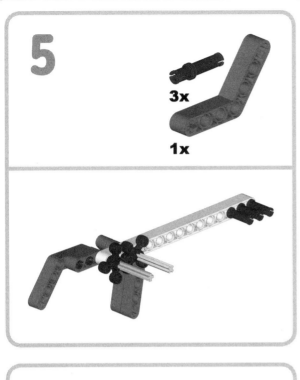

3x

1x

6

2x

You will have to turn the blue peg to make it easier to add the angled beam in step 7.

7

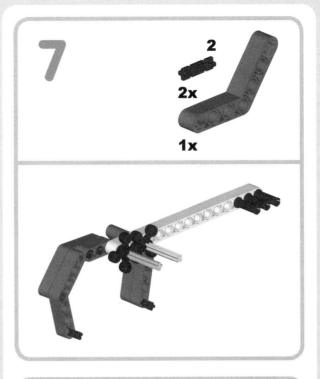

2

2x

1x

8

2x

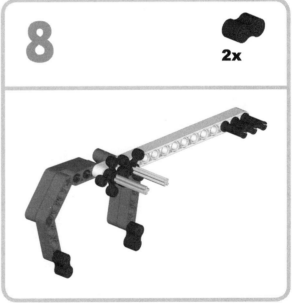

9

1x

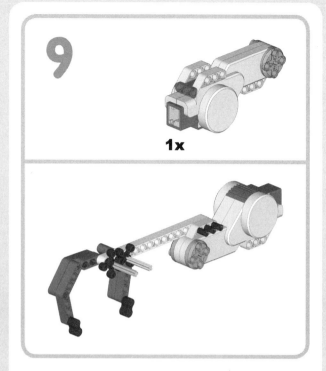

10

2x

1x

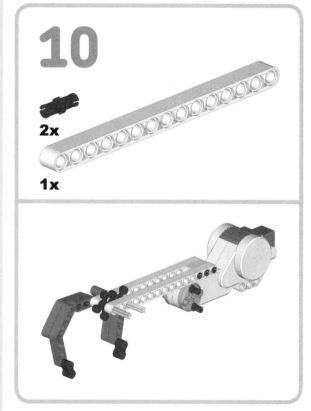

Use light gray pegs (non-friction) in steps 11 and 12.

11

1x

1x

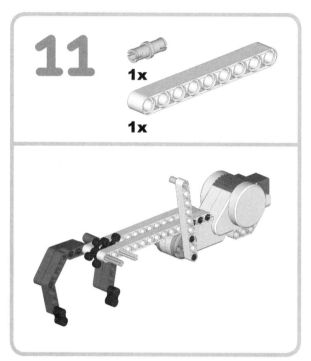

12

1x

1x

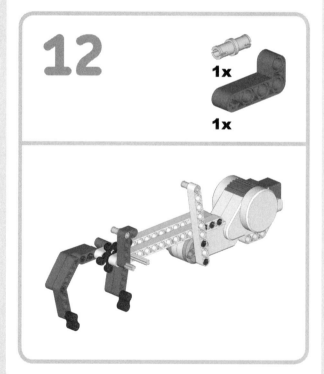

13

1x

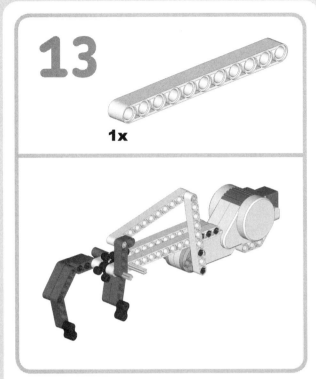

14

3x

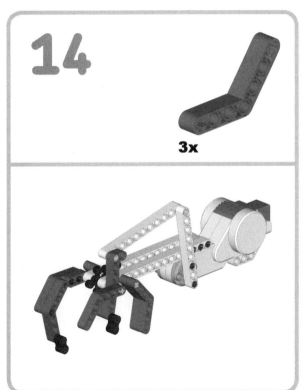

15

2
2x
2x
1x

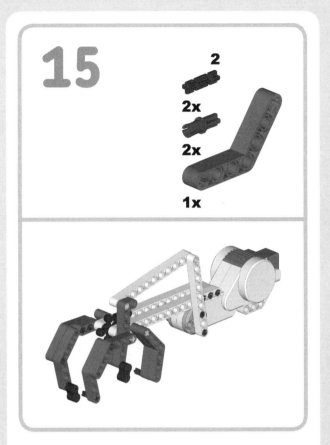

16

2x

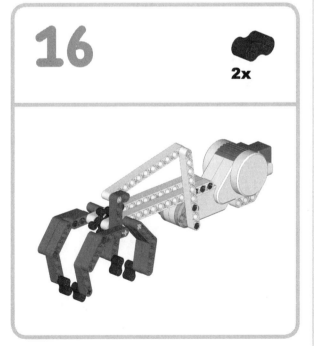

Finally, use the longest NXT cable to connect the base motor to the gripper motor, and you are ready to have some fun!

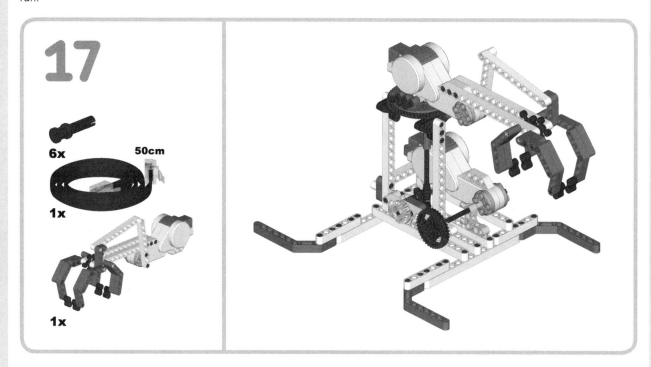

program-ming the CandyPicker

As mentioned at the beginning of this chapter, the Candy-Picker requires no programming. Grab some candy and put your new creation to work!

further exploration

You will notice that the CandyPicker's arm cannot bend at its shoulder joint. As a result, you must place the candy at a fixed height in order for the robot to work, since the it cannot reach down.

Your challenge is to modify the design and lever arrangements so that the arm has another "degree of

freedom"; in other words, modify the CandyPicker so that its arm can swing up and down and it can open and close the gripper at the same time.

Figure 1-3 shows a sample modification to get you started.

Figure 1-3: A modified version of the CandyPicker

2

PunchBot: old-school programming using your NXT

Figure 2-1: PunchBot

Computers haven't always had a mouse or keyboard available to use for programming. Long before mice and keyboards became standard tools, one typical method for inputting data was the simple punch card (also known as a *Hollerith card*, named for its creator, Herman Hollerith).

Punch cards came in a variety of size and shapes, but what they all had in common was how they were used. Square or rectangular holes were punched in the cards using a special tool. You've probably heard that computers "talk" using 1s

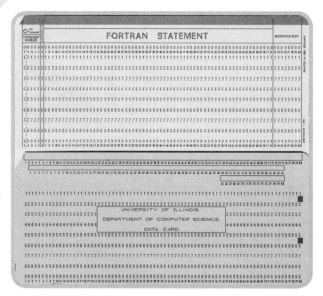

Figure 2-2: Two sample punch cards

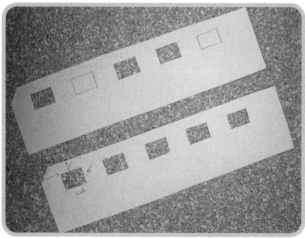

Figure 2-3: PunchBot's punch cards look like these.

and 0s; another term for this is *binary*. Binary programming relies on a computer being able to distinguish a 1 from a 0. It may seem hard to believe, but computers today still communicate using a series of 1s and 0s.

NOTE Learning binary is not difficult; if you would like to learn more about binary and how computers and robots use binary to communicate, please visit the reference websites listed in "Internet Resources" on page 47.

In this chapter, you'll build a robot that will simulate how computers were programmed decades ago. By "feeding" your robot a series of punch cards, you will be able to program the robot to perform up to 16 different activities (and you can even create the activities the robot will perform).

How does this work? Each punch card you make will have a series of five *bits*. Each bit is represented by a square on the punch card, and the squares are spaced in a straight line. When a bit is cut out (also known as open or punched), it represents a 1. When the bit is not cut out (closed or unpunched), it represents a 0. The first bit on the card is known as the *synchronization bit*. This bit must always be open and will allow the robot to properly read the remaining four bits on the card (the robot's program knows how far to move the card forward to read each subsequent bit). Bits are read left to right, so the card at the top of Figure 2-3 would be read as 1 0 1 1 0 (open-closed-open-open-closed). Remember that the first bit is the synchronization bit, so we'll ignore it and write the sequence of bits as 0 1 1 0.

Below are all the possible configurations for the remaining four bits.

0	0	0	0
0	0	0	1
0	0	1	0
0	0	1	1
0	1	0	0
0	1	0	1
0	1	1	0
0	1	1	1
1	0	0	0
1	0	0	1
1	0	1	0
1	0	1	1
1	1	0	0
1	1	0	1
1	1	1	0
1	1	1	1

Count them. You should find that there are 16 different configurations, ranging from 0000 (closed-closed-closed-closed) to 1111 (open-open-open-open). These match with 16 different actions that your robot will be able to perform, all based on the order of the cards you insert.

As an example, you might choose to assign the configuration of 1010 (open-closed-open-closed) to a robot action of "Move forward three rotations and turn left 90 degrees." Likewise, configuration 0101 might correspond to "Move backward three rotations and turn right 90 degrees."

NOTE You don't have to decide on all the robot actions at this time, but start thinking about what you'd like your robot to be able to do; when you create the NXT-G program later in the chapter, you will see where you will drop in the additional programming blocks required for each of the 16 actions. Each possible card sequence corresponds to a special action (or actions) that you will program the robot to perform. For example, you might choose to give a card sequence of 1 0 1 0 a corresponding action of "Move forward 5 rotations."

After building and programming PunchBot using punch cards, you'll probably have a greater appreciation for people who programmed robots in the early days of computing. It was time consuming, as you'll discover. You'll cut out one punch card, fifty, or more if you like. You'll then have to "punch" the cards to create your own binary program. This binary program will be the set of instructions telling PunchBot which actions to take.

When you're done, you'll proudly be able to say that you've done some simple binary programming. Have fun!

building PunchBot

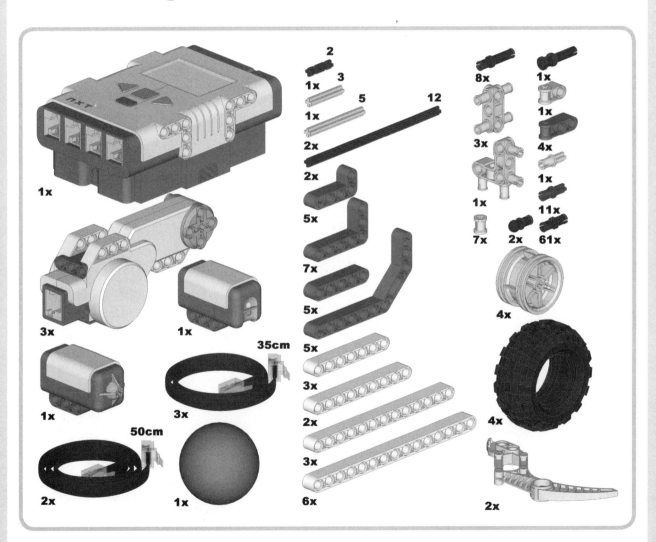

Figure 2-4: Bill of Materials

the main body

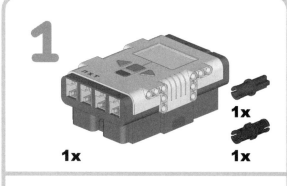

1

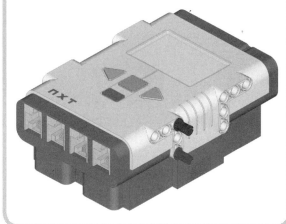

1x

1x

1x

2

1x 1x

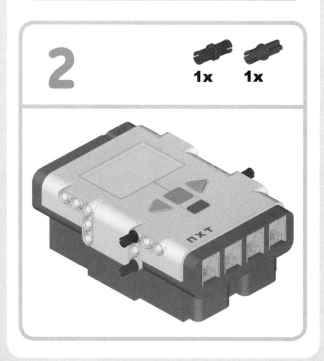

3

3x

1x

2x

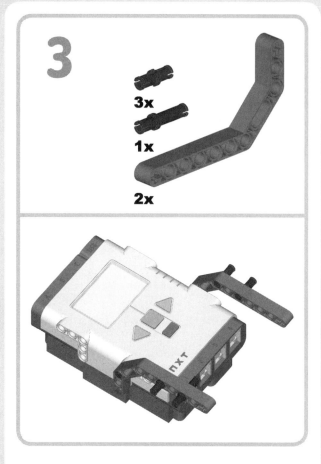

4

1x

1x 1x

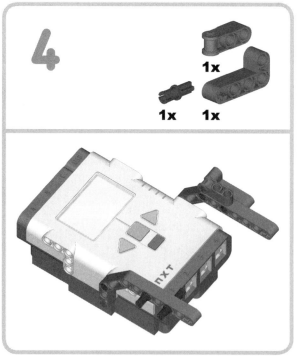

5

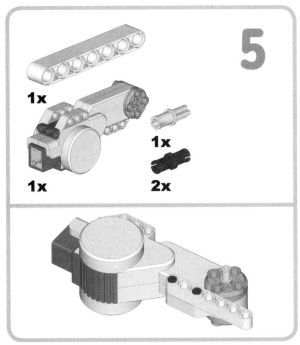

1x

1x

1x

2x

6

1x

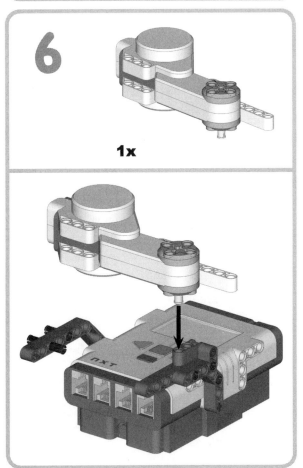

7

2x

1x

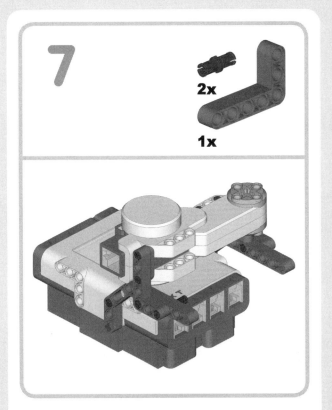

8

2x

1x

1x

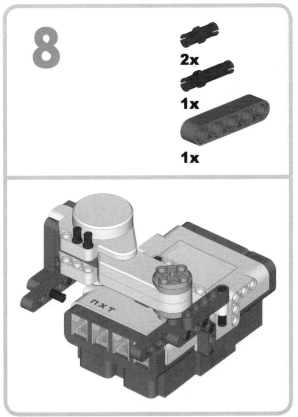

9

1x
1x
1x

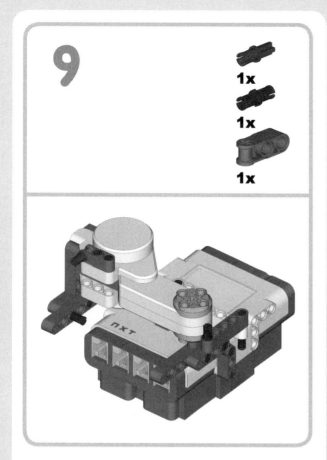

3x
1x

10

1x

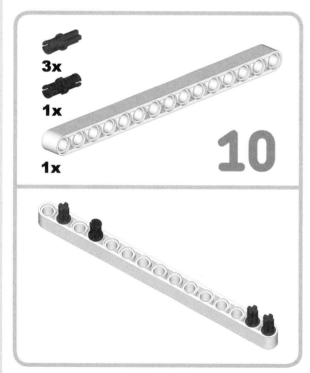

11

2x
1x
1x

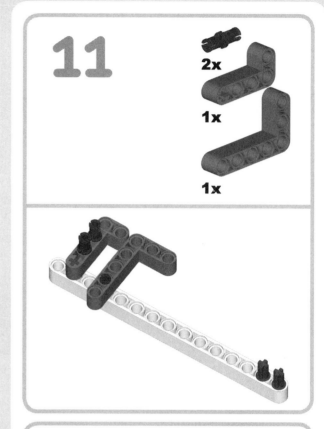

12

1x
1x
1x

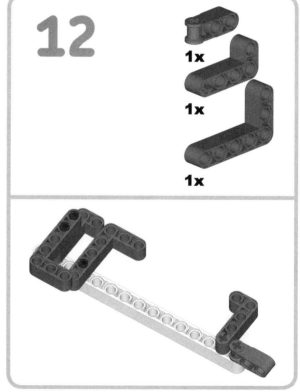

13

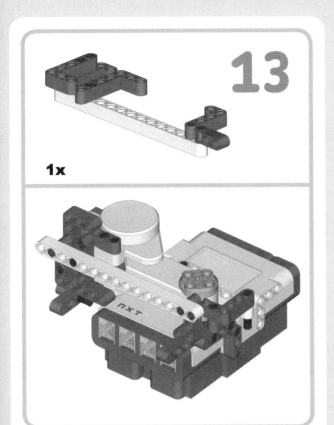

1x

14

2x

1x

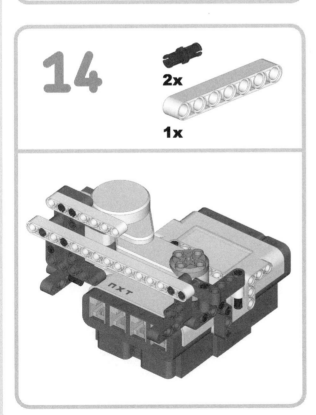

15

1x

1x

16

6x   1x

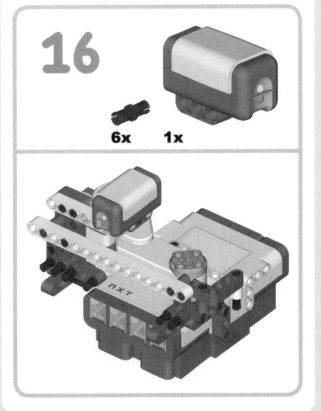

17

2x
1x

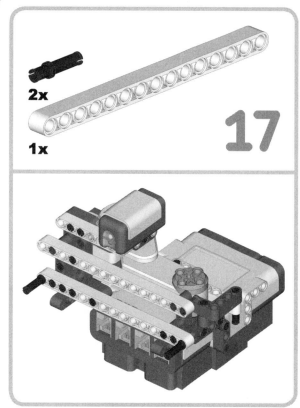

18

4x
2x

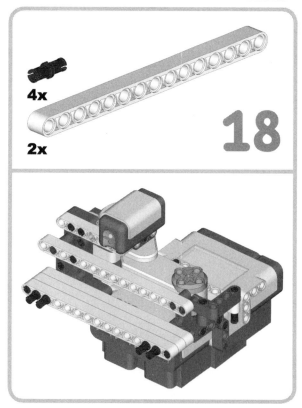

19

1x
2x
3
2x
1x

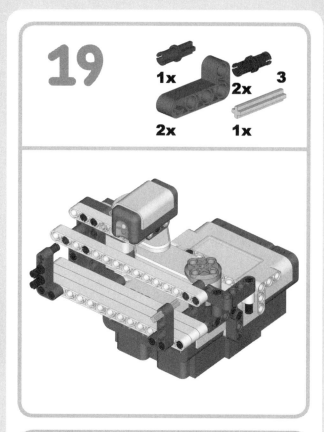

20

1x
1x
1x
1x

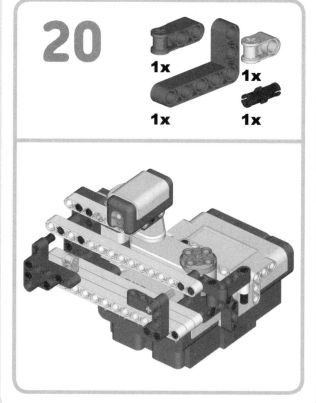

21

2

1x

1x

1x

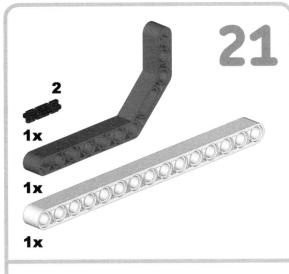

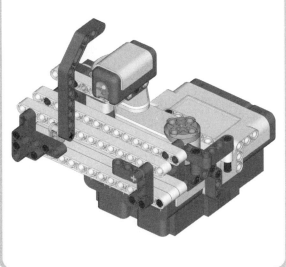

22

2x

12

3x

2x

23

24

2x

2x

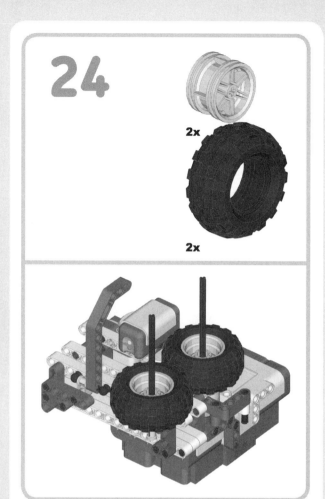

26

1x

25

2x

27

1x

1x

1x

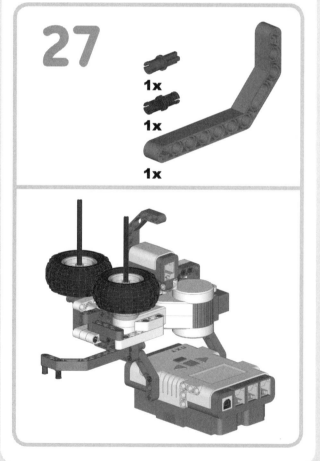

28

1x
1x
1x

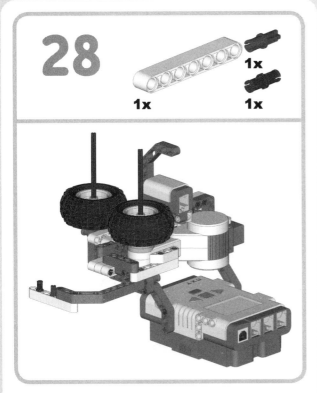

29

1x

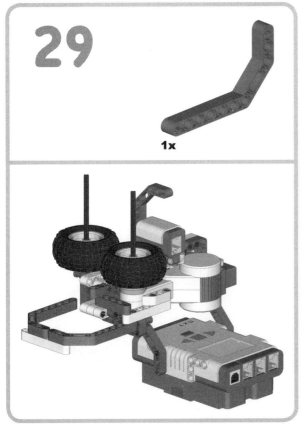

30

1x
1x

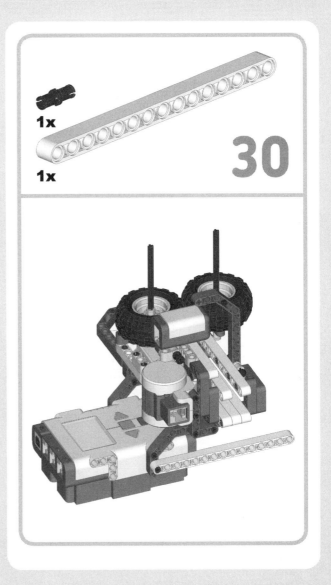

31

4x

1x

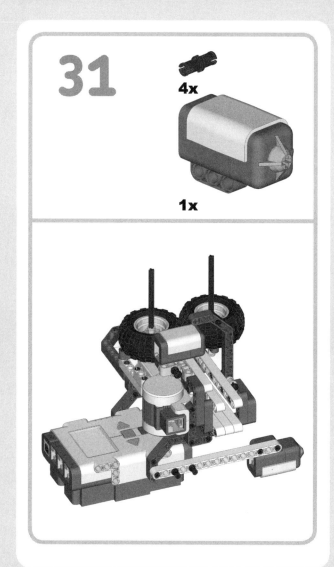

32

1x

1x

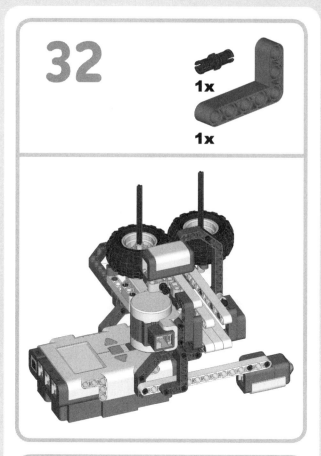

33

2x

the motor and wheels

1

4x

2x

1x

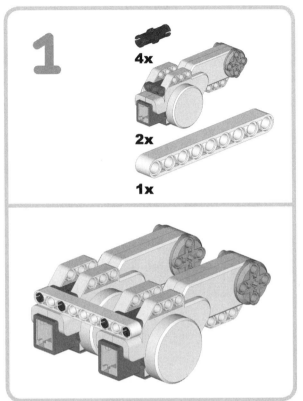

2

2x

5

2x

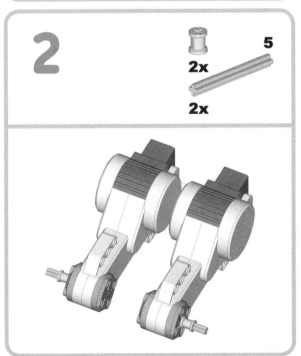

3

2x

2x

2x

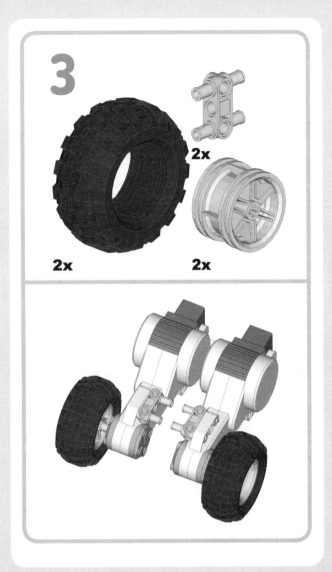

4

4x

2x

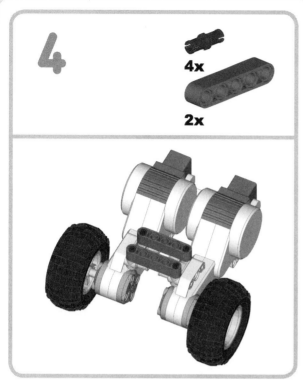

5

1x

6

4x

1x

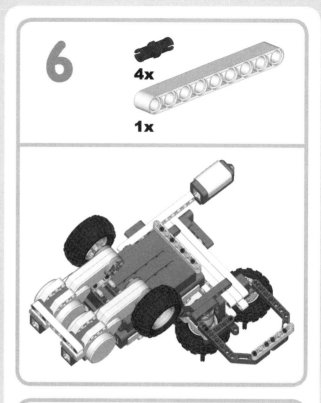

7

3x

1x

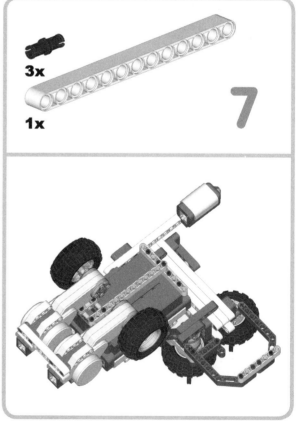

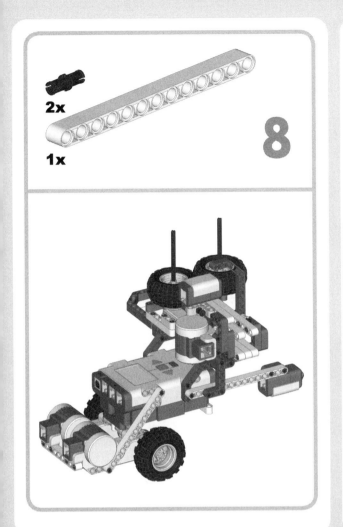

8

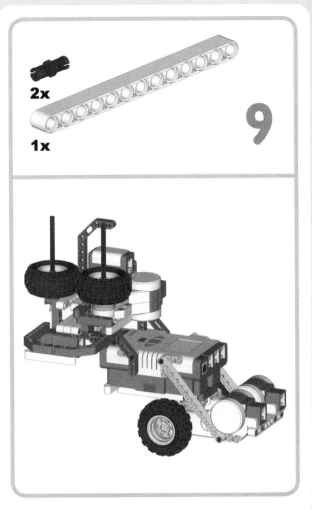

9

the caster

1

2x

2

4x

2x   1x

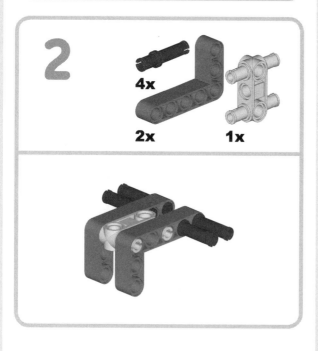

3

2x   2x

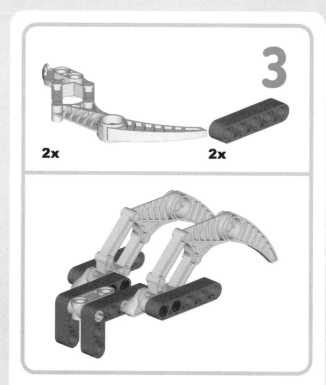

4

1x

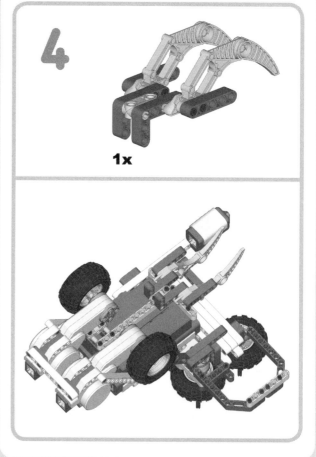

5

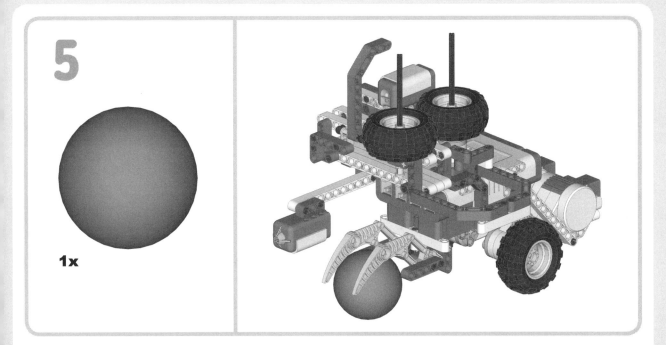

1x

connecting the cables

1

35cm

50cm

3x

2x

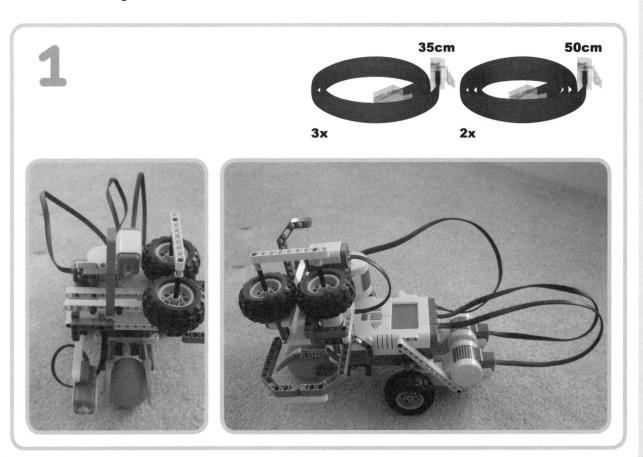

punch card instructions

After you've built the robot, you'll also need to create some punch cards. Create these from poster board or a medium to heavy card stock. The dimensions for the punch cards are provided in Figure 2-5.

NOTE Did you notice the clipped edge on the punch card in the upper-left corner? This is useful for keeping your punch cards oriented properly. If you stack your punch cards top to bottom in the order you want the robot to perform the actions, the clipped edge will help to ensure that none of your cards are upside down or flipped over. If your stack doesn't have all the clipped edges in the upper-left corner, you'll know that one or more cards in your stack are inserted improperly.

programming PunchBot

The program for PunchBot contains comments that can help you understand each programming block's function. Additional comments are provided for various sections of the program.

The program is broken into sections of four or five programming blocks, with each programming block's configuration panel provided.

For simplicity, it will be easier if you drag and drop two Loop blocks onto the work space and then place two additional Loop blocks inside the first Outer Loop block, as shown in Figure 2-6. Configuration panels for each Loop block are also provided (Loop block 1 is at the top, Loop block 2 is second, Loop block 3 is third, and Loop block 4 is at the bottom).

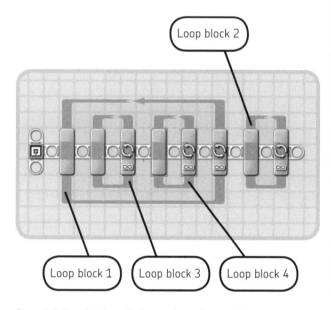

Figure 2-6: Drop four Loop blocks onto the work space before adding any additional blocks.

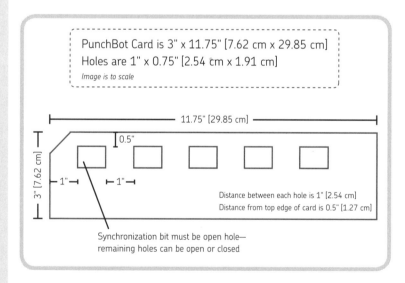

PunchBot Card is 3" x 11.75" [7.62 cm x 29.85 cm]
Holes are 1" x 0.75" [2.54 cm x 1.91 cm]
Image is to scale

11.75" [29.85 cm]

0.5"

3" [7.62 cm]

1"   1"

Distance between each hole is 1" [2.54 cm]
Distance from top edge of card is 0.5" [1.27 cm]

Synchronization bit must be open hole—
remaining holes can be open or closed

Figure 2-5: Dimensions for a PunchBot punch card

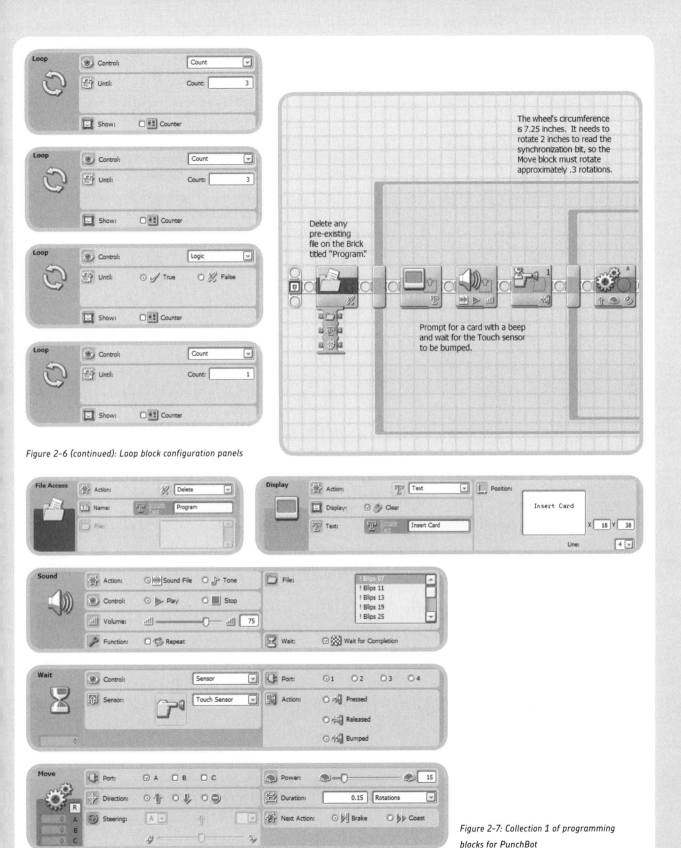

Figure 2-6 (continued): Loop block configuration panels

The wheel's circumference is 7.25 inches. It needs to rotate 2 inches to read the synchronization bit, so the Move block must rotate approximately .3 rotations.

Delete any pre-existing file on the Brick titled "Program."

Prompt for a card with a beep and wait for the Touch sensor to be bumped.

Figure 2-7: Collection 1 of programming blocks for PunchBot

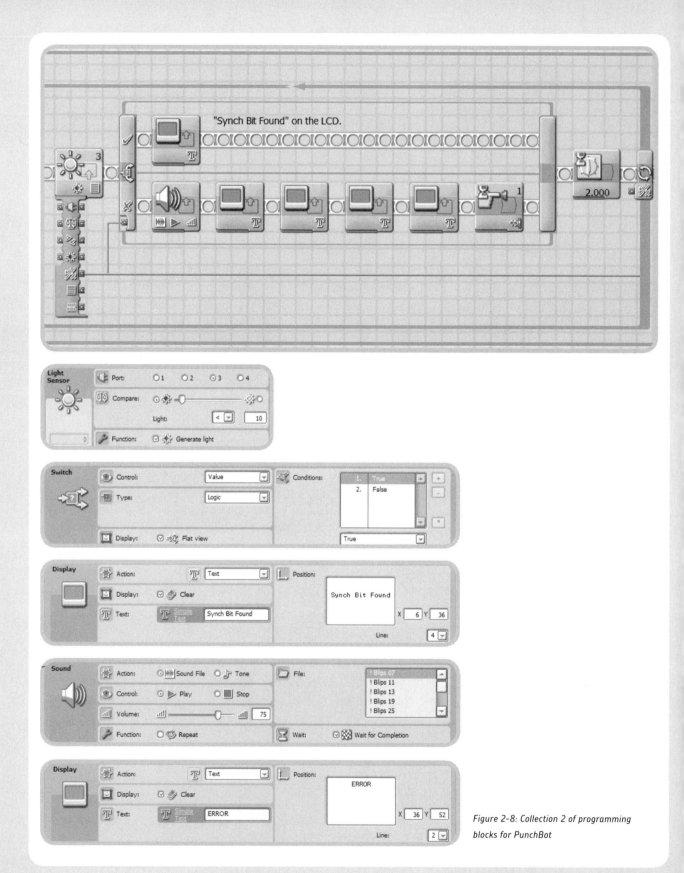

"Synch Bit Found" on the LCD.

2.000

Light Sensor

Port: ○1 ○2 ⊙3 ○4

Compare: ⊙ ☀ ——— ☀ ○

Light: < ▾   10

Function: ☑ ☀ Generate light

0

Switch

Control: Value ▾

Type: Logic ▾

Display: ☑ ⚡ Flat view

Conditions:
| 1. | True |
| 2. | False |

True ▾

Display

Action: 𝕋 Text ▾

Display: ☑ ◇ Clear

Text: 𝕋 Simple Text | Synch Bit Found

Position:

Synch Bit Found

X 6 Y 36

Line: 4 ▾

Sound

Action: ⊙ Sound File ○ ♪ Tone

Control: ⊙ ▷ Play ○ ■ Stop

Volume: ——— 75

Function: ○ ⟳ Repeat

File:
! Blips 07
! Blips 11
! Blips 13
! Blips 19
! Blips 25

Wait: ☑ Wait for Completion

Display

Action: 𝕋 Text ▾

Display: ☑ ◇ Clear

Text: 𝕋 Simple Text | ERROR

Position:

ERROR

X 36 Y 52

Line: 2 ▾

Figure 2-8: Collection 2 of programming blocks for PunchBot

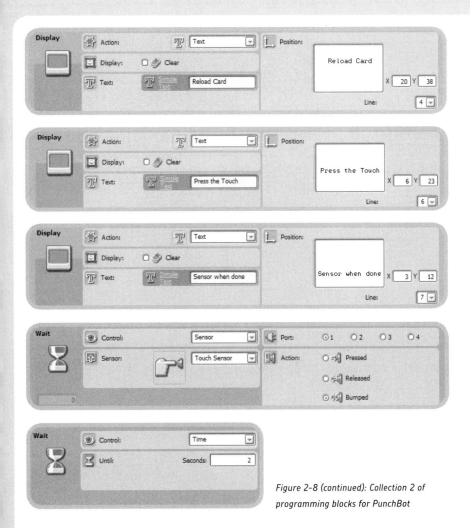

Figure 2-8 (continued): Collection 2 of programming blocks for PunchBot

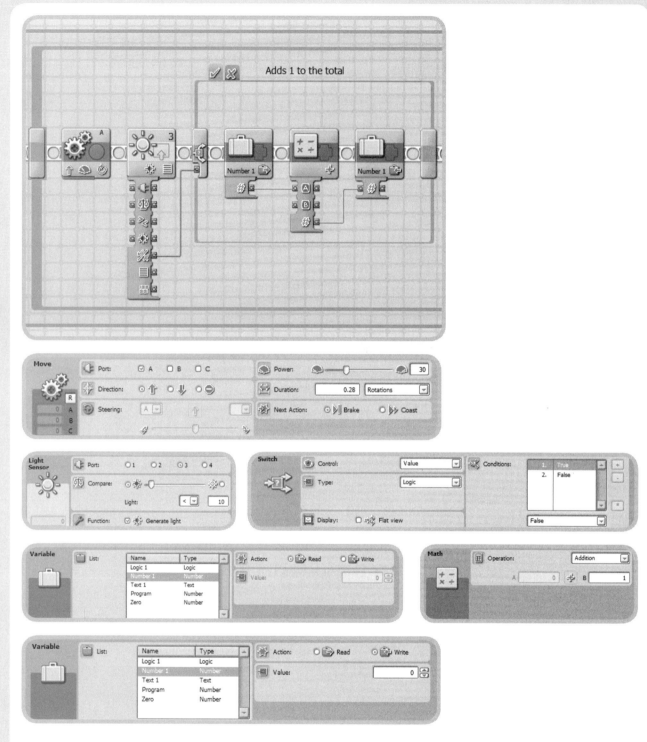

Figure 2-9: Collection 3 of programming blocks for PunchBot

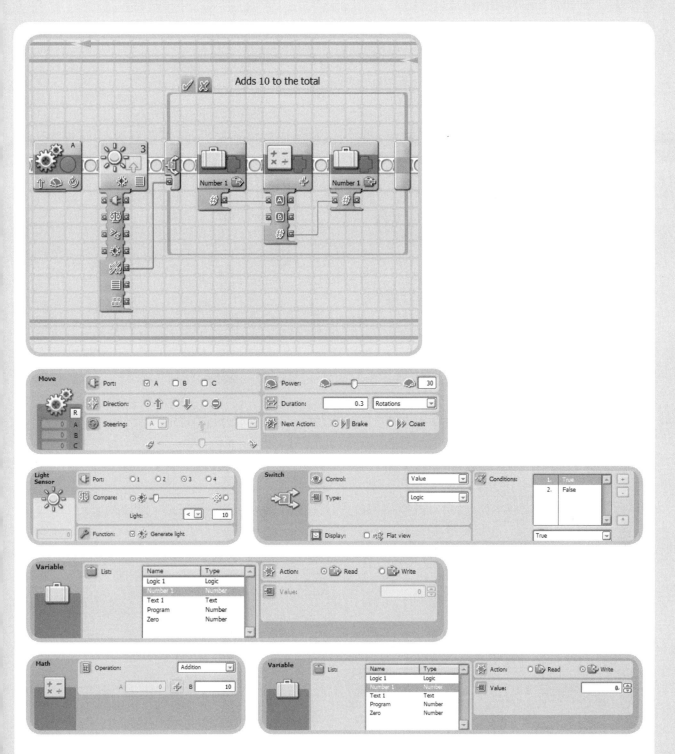

Figure 2-10: Collection 4 of programming blocks for PunchBot

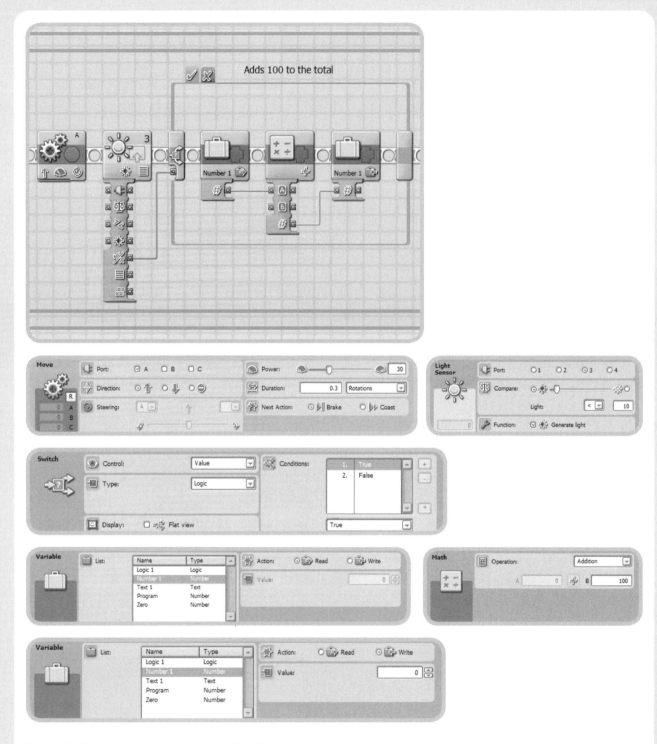

Figure 2-11: Collection 5 of programming blocks for PunchBot

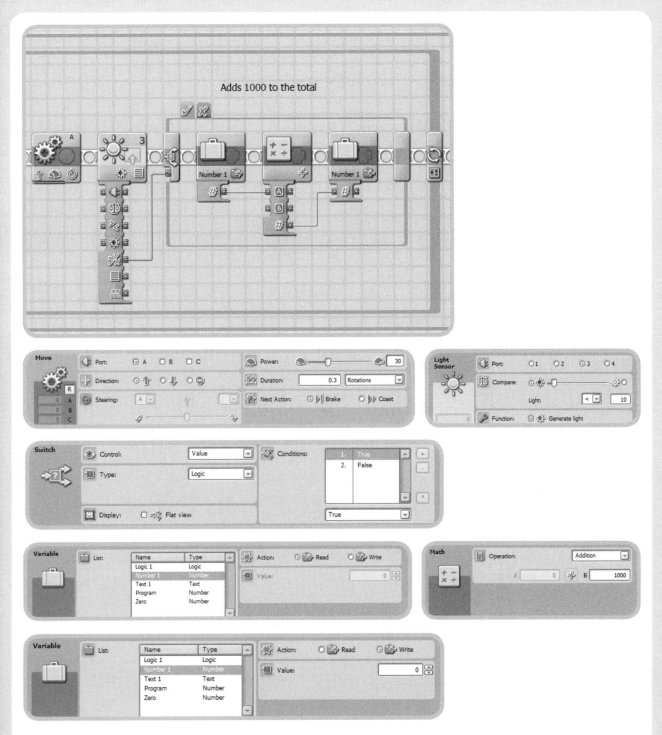

Figure 2-12: Collection 6 of programming blocks for PunchBot

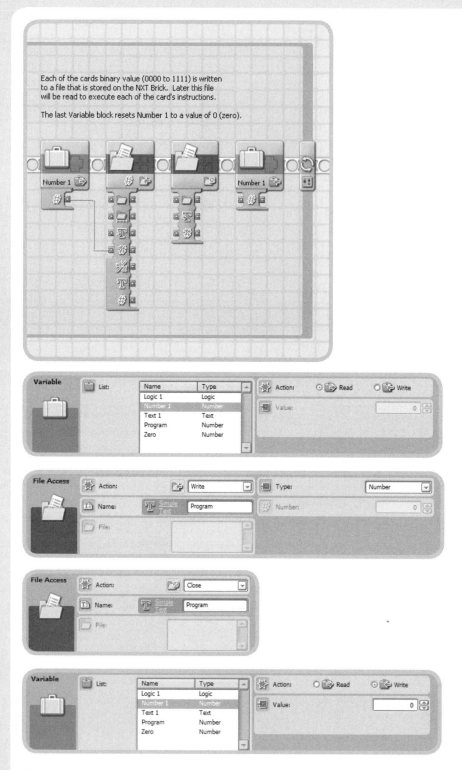

Figure 2-13: Collection 7 of programming blocks for PunchBot

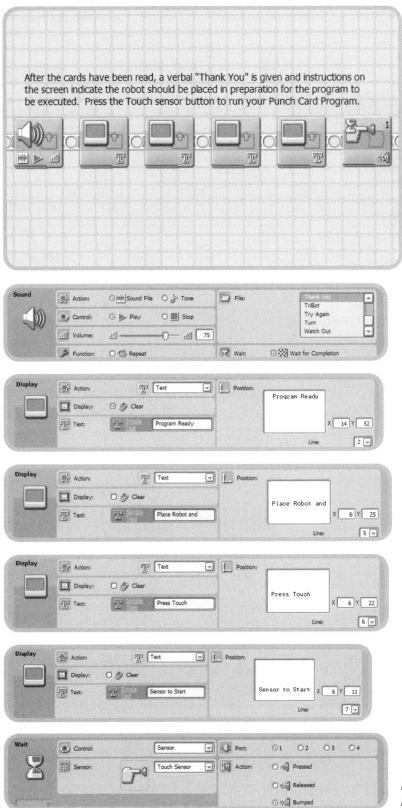

After the cards have been read, a verbal "Thank You" is given and instructions on the screen indicate the robot should be placed in preparation for the program to be executed. Press the Touch sensor button to run your Punch Card Program.

Figure 2-14: Collection 8 of programming blocks for PunchBot

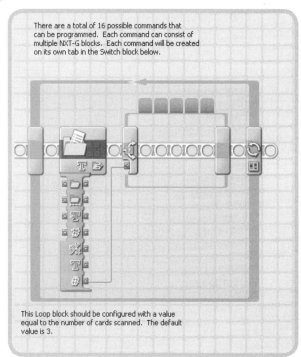

There are a total of 16 possible commands that can be programmed. Each command can consist of multiple NXT-G blocks. Each command will be created on its own tab in the Switch block below.

This Loop block should be configured with a value equal to the number of cards scanned. The default value is 3.

There are 16 configurable tabs on the Switch block shown in Figure 2-15 (only 5 are visible, so click the number of the tab you wish to use in the Conditions section of the Switch block's configuration panel). A Switch block typically has two available paths (or options) for the programmer to use. But in a situation like this one, the robot will have to make a selection from more than two choices. To do this, you must configure the Switch block to be displayed in a tabbed view. Flat view is the default view for the Switch block when placed in a program. To change to a tabbed view, uncheck the box labeled *Flat view*, and the Switch block will now display two tabs. Consult the Switch block Help documentation for instructions on adding additional tabs.

There are a total of 16 options for PunchBot, one for each of the possible configurations of a punch card. In Figure 2-15, make note of the values for each of the 16 options and their order (0, 1, 10, 11, 100, 101, 110, 111, 1000, 1001, 1010, 1011, 1100, 1101, 1110, 1111).

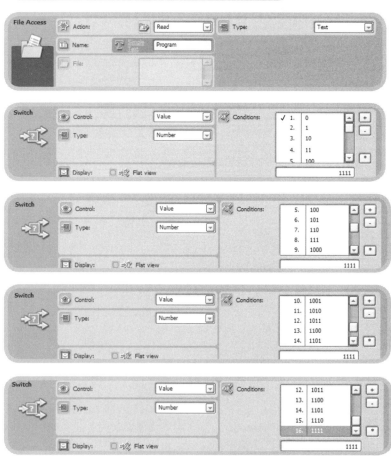

Figure 2-15: Collection 9 of programming blocks for PunchBot

For each of the 16 tabs in the Switch block, you can drag and drop additional NXT-G blocks to create PunchBot's 16 programmable actions. If NXT-G blocks are *not* placed in one or more tabs, any punch cards corresponding to those values will have no actions for PunchBot to perform.

further exploration

Now that you have built PunchBot and created the program for it, spend some time experimenting by adding more NXT-G blocks inside the last Switch block's tabs. You can start simply, for example, by giving a card with the value of 0000 the assigned actions of moving your robot forward three rotations. You might then give the next value, 0001, the reverse action of moving the robot backward three rotations. Next, assign values 0010 and 0011 the actions of rotating right 90 degrees and left 90 degrees, respectively. After programming these four actions, you still have 12 more programmable actions you can give your robot. After you've finished providing your robot with 16 unique actions, here are some suggested activities you can do with your new PunchBot:

* Play a game with your friends. Team 1 secretly programs the robot with a set of cards. After programming the PunchBot using the cards, shuffle the cards and give them to Team 2. Let Team 2 observe the robot's actions and then try to put the punch cards in the proper order that represents the robot's actions.
* Have Team 1 set up an obstacle course on the floor. Team 2 will use the punch cards to program the robot to navigate the obstacle course. Change the obstacle course and let Team 1 try to program the robot to navigate Team 2's obstacle course.

modifications to PunchBot

Here are some suggestions for modifying or improving PunchBot:

* If a rechargeable battery is available, modify PunchBot's design to allow for the use of the battery. This may require a redesign of the ball caster and/or relocation of the wheels.
* Consider modifying the punch card reader to use the Ultrasonic Sensor instead of the Light Sensor. Does the Ultrasonic Sensor work better or worse than the Light Sensor?

modifications to PunchBot's program

Here are some suggestions for modifying or improving PunchBot's program:

* Instead of PunchBot moving with its motors and wheels, modify the robot to perform simple calculations such as two plus two (2 + 2). Using punch card 1 and punch card 3, how would you change the program to allow the four bits to hold a value such as 1 or 5 or 15? How would you modify the program so that punch card 2 determines whether the number assigned by punch card 1 is added, subtracted, multiplied, or divided by the number assigned by punch card 3?
* If you modify the punch cards and add a fifth bit (six bits in all, counting the synchronization bit), how many more possible actions would be possible? How would you modify the program to allow for this additional bit?

internet resources

The history of computers is an interesting one. It's surprising that early programmers had to punch small holes in cards to tell computers what to do. Computing has come a long way to reach the graphical drag-and-drop programming method found in a simple NXT-G program!

If you've enjoyed learning a little bit about binary programming and wish to learn more, visit the excellent websites in the following list. Have fun!

* *http://computer.howstuffworks.com/bytes.htm*
* *http://www.garlikov.com/Soc_Meth.html*
* *http://www.instructables.com/id/Binary-Counting/*
* *http://homepage.ntlworld.com/interactive/*
* *http://www.computerhistory.org/timeline/*
* *http://www.computersciencelab.com/ComputerHistory/History.htm*

3

m: the m&m sorter

Every day we carry out monotonous tasks that we wish someone or something else would help with. In fact, early machines (like the water mill shown in Figure 3-2) were designed to free people of simple repetitive tasks, such as grinding wheat or drawing water from a well.

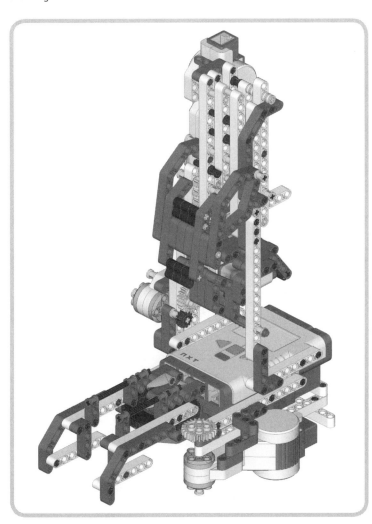

Over time, machines became increasingly complex and performed more ambitious functions, like transporting people, weaving cloth (like the mechanical loom shown in Figure 3-3), and cleaning clothes.

As time passed, technology became more sophisticated, but machines still required human supervision. When we began to control our machines with internal instructions—called *programs*—we were able to leave our equipment to perform its work autonomously. These new machines could respond appropriately to changes in the environment on their own, and we now know them as robots.

Today, robots make it possible for us to manufacture items like cars, planes, and computers on a large scale that was unimaginable in the past, but they still serve their original purpose—freeing us from tedious chores. Service robots in the home are expected to be a primary growth market in the 21st century. Now that you have the NXT kit, there's no need to wait any longer—you can begin designing your own service robots.

This chapter will show you how to build and program an NXT-based robot that can sort peanut M&M candies by color. The more-or-less regular and round form of the peanut M&M makes it perfect for our needs.

Figure 3-1: M—the M&M sorter

Figure 3-2: Water mill

When you've finished building, programming, and testing your M&M sorter, you will have a service robot at your disposal that relieves you from a truly tedious and mechanical task (tongue firmly in cheek). You will know how to use Light Sensors, how to design robots for particular tasks (like storing, scanning, and sorting items in a warehouse), and how to control a sequential work flow.

We will use different subcomponents for each of these tasks, as shown in Figure 3-4 and described in the following paragraphs. The parts themselves should be pretty self-explanatory.

The robot's store is the place where we initially put the M&Ms that need to be sorted. M&Ms are dropped one at a time from the store into the scanner unit. The store's cover prevents the candy from overshooting and missing the unit.

In the scanner unit, which is driven by a single motor, the M&M is checked for its color. A Light Sensor measures the light reflected from the candy, and the program deduces its color from this measurement.

Subsequently, the collector is adjusted accordingly by another motor. Depending on the color value computed, it switches horizontally to a certain angle.

When the collector is aligned, the M&M is dropped from the scanner unit, and it makes its way through the collector.

Finally, all components are reset to their initial state, and the next M&M is processed.

Figure 3-3: Mechanical loom

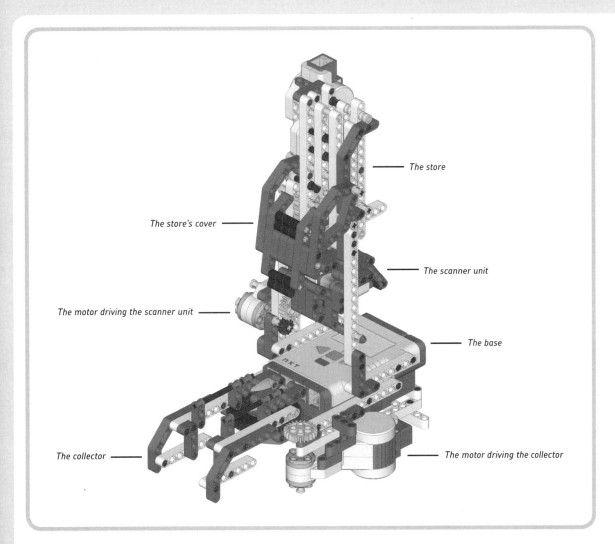

The store

The scanner unit

The store's cover

The motor driving the scanner unit

The base

The collector

The motor driving the collector

Figure 3-4: Overview of the different components

building m

We will build M by first building the different components and then assembling them as the complete robot.

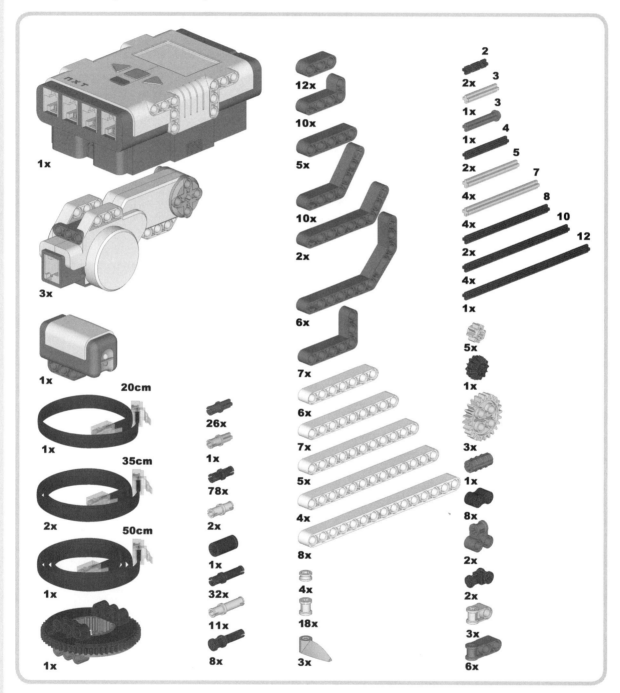

Figure 3-5: Bill of Materials

the brick

1

1x 3x 2x

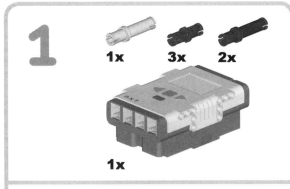

1x

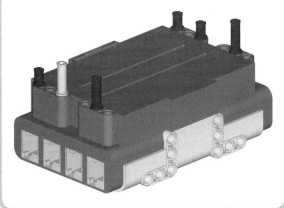

2

2x 2x

2x 1x

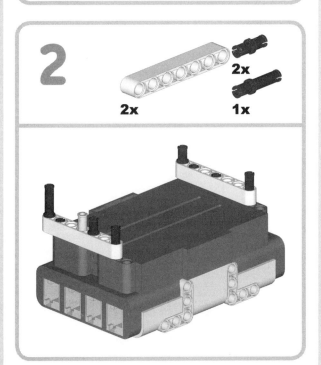

3

4x

2x

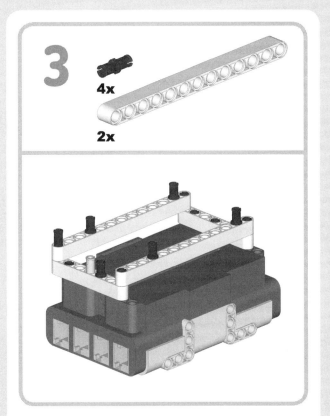

4

1x

2x

1x

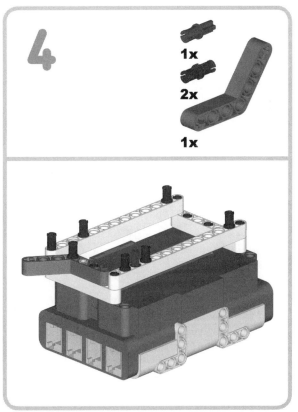

5

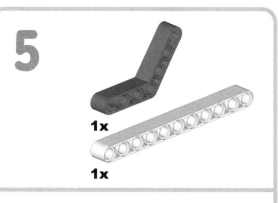

1x

1x

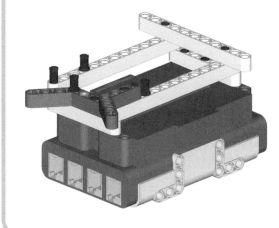

6

8x

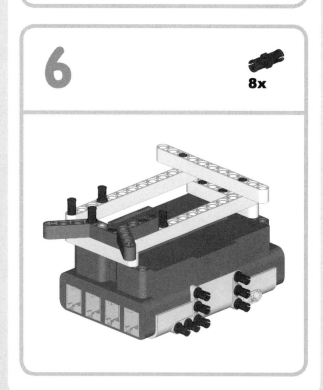

the motor driving the collector

1

1x 1x

1x

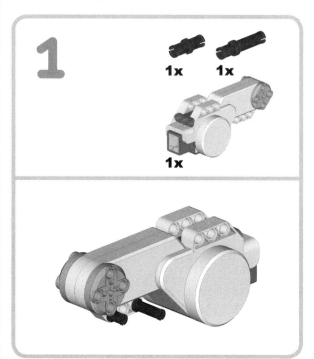

2

1x

1x

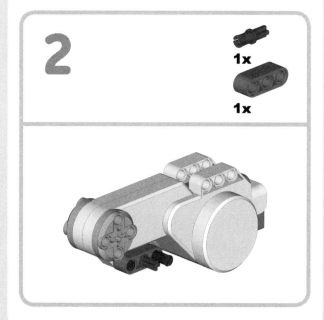

3

2x **1x**

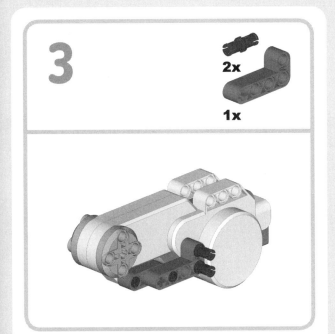

4

4x **1x** **1x**

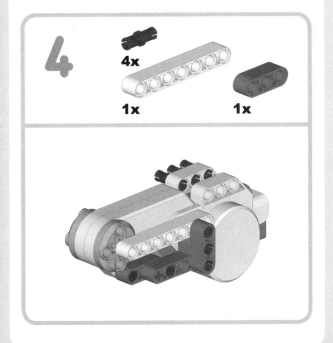

5

1x **3x** **1x**

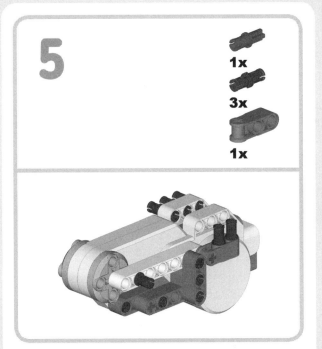

6

1x **1x**

7

1x

7

1x

1x

3x

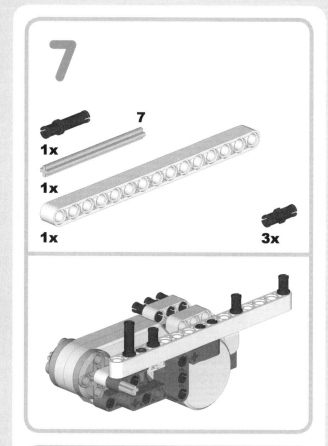

8

1x

1x   1x   1x

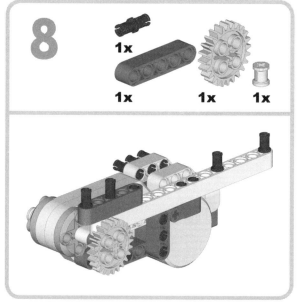

9

1x   1x

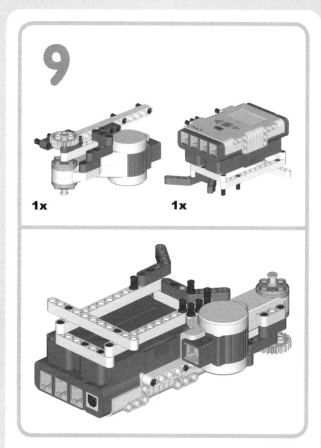

10

1x

1x

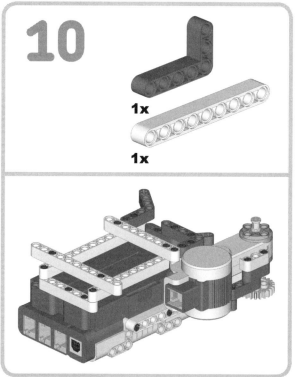

the motor driving the scanner unit

1

4x   1x

1x

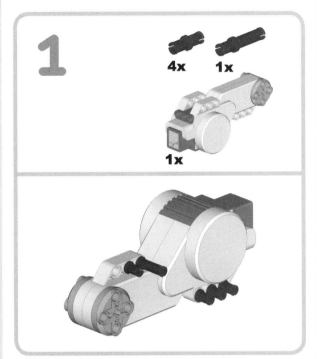

2

2x

1x

1x

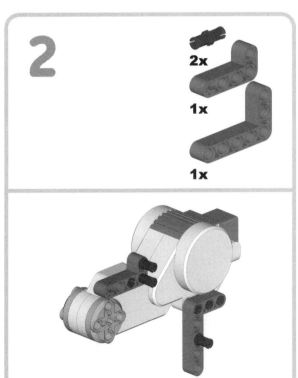

3

1x

1x

4

1x

2x

1x

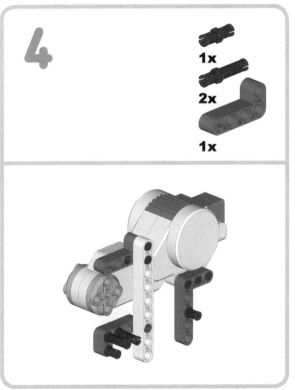

5

2x
1x

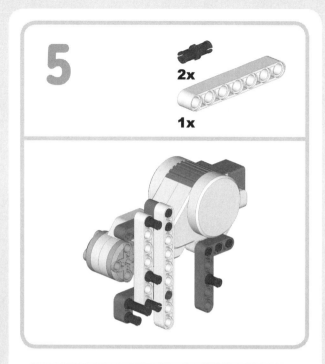

6

2x
7 1x
12 1x
1x

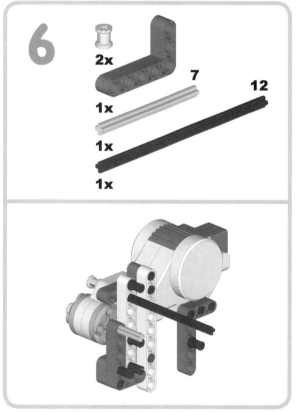

7

1x
1x

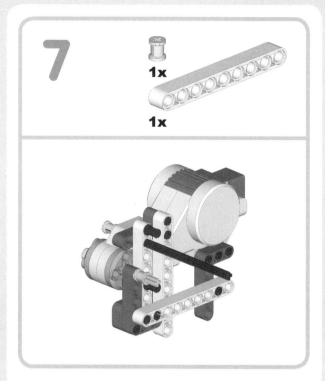

8

2x
1x

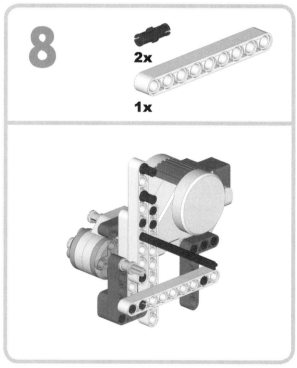

9

1x

10

1x

1x

1x

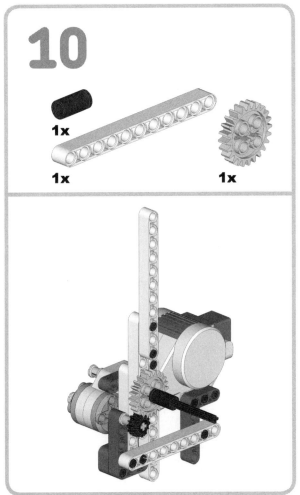

11

2x

4

2x

2x

2x

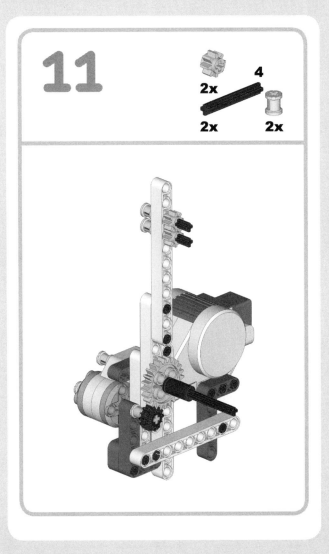

12

1x

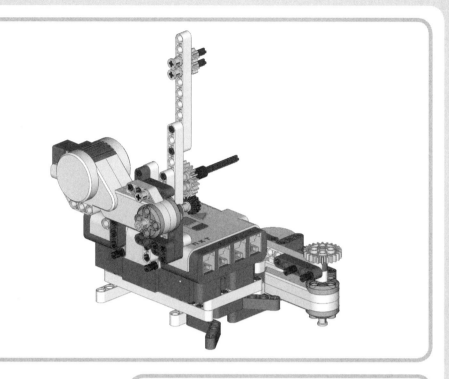

the collector

1

2x

1x

2

2x

1x

3

1x

4

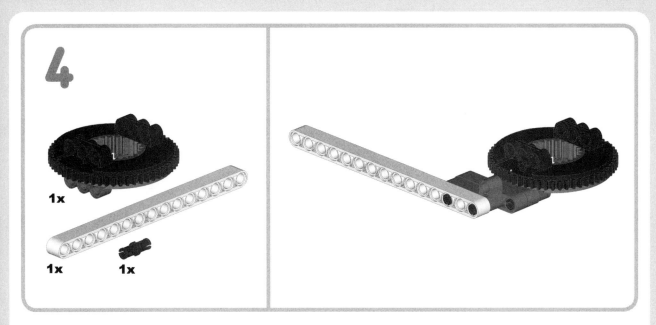

1x

1x 1x

5

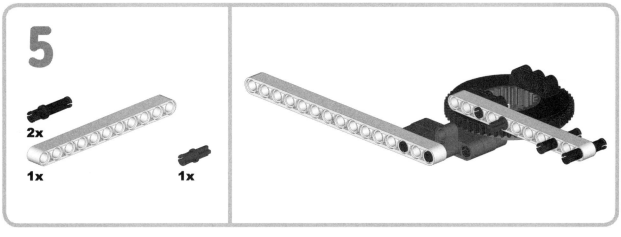

2x

1x 1x

6

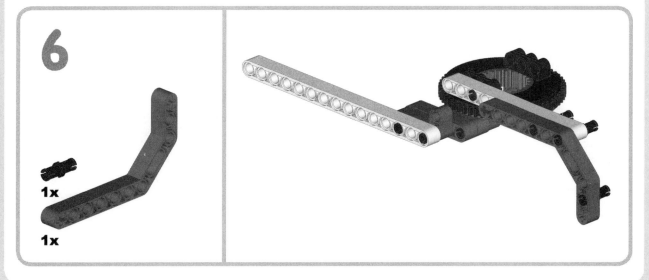

1x

1x

7

1x   1x

2x

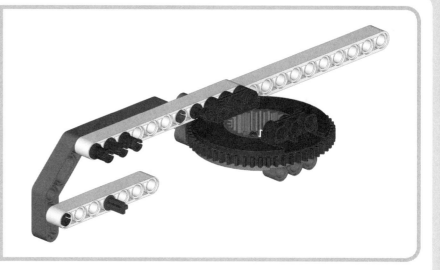

8

2x

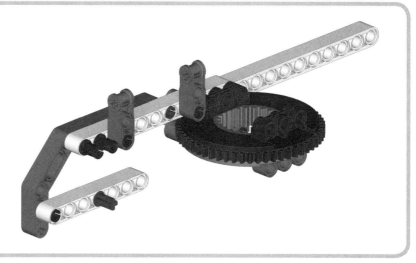

9

1x

7

3x

1x   1x

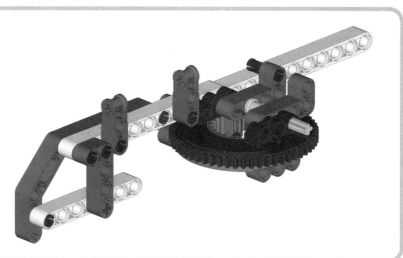

10

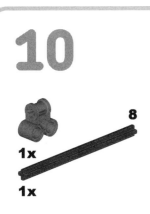

1x

8

1x

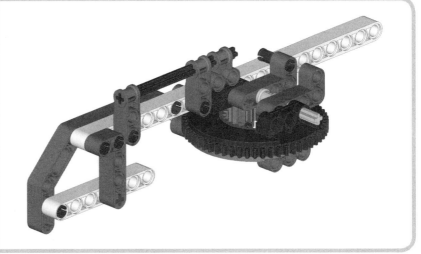

11

1x

1x

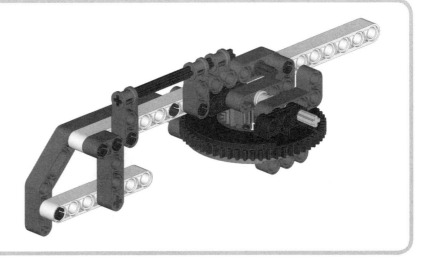

12

2x

7

1x

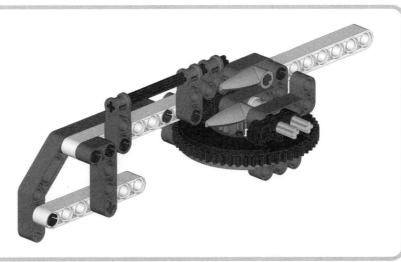

13

2x

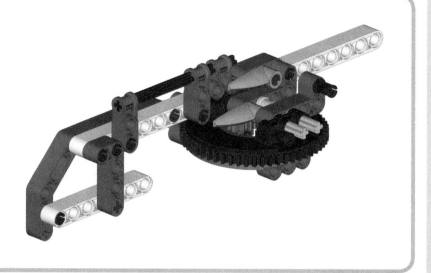

14

1x   1x   1x

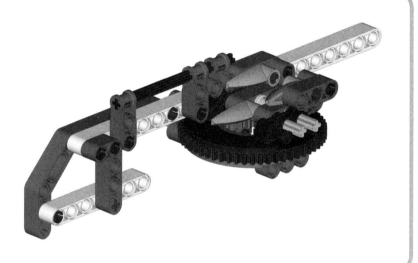

15

1x

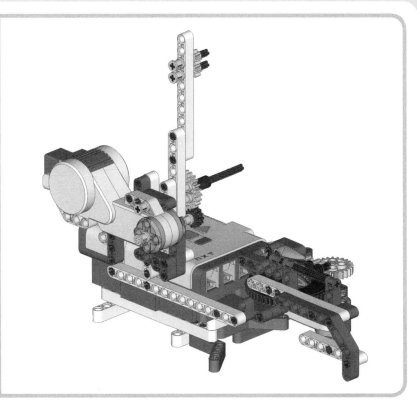

16

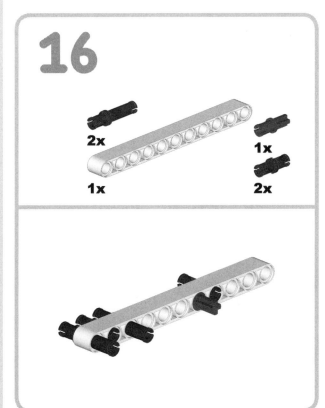

2x

1x

1x

2x

17

1x

2x

1x

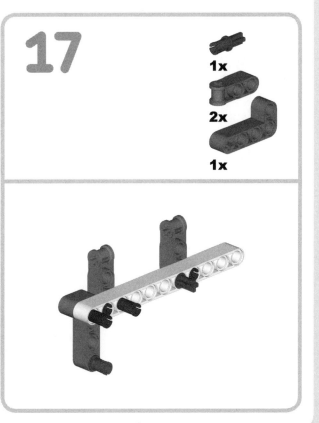

18

1x
1x

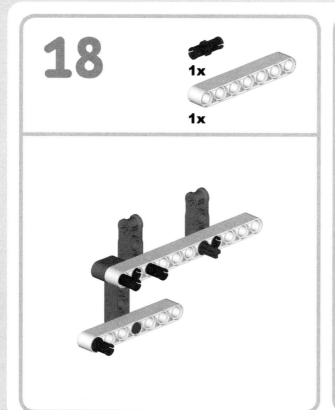

19

8

1x
1x
1x

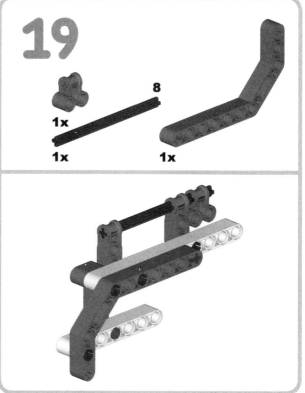

20

1x

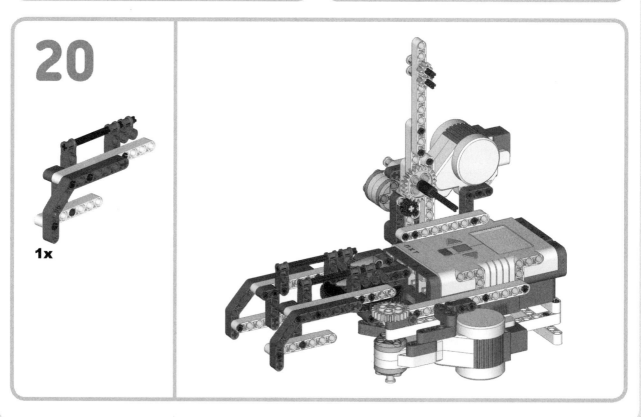

the scanner unit

1

2x

1x

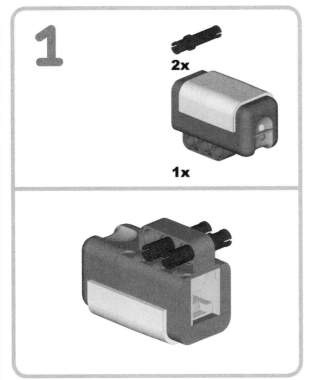

2

2x

4x

2x

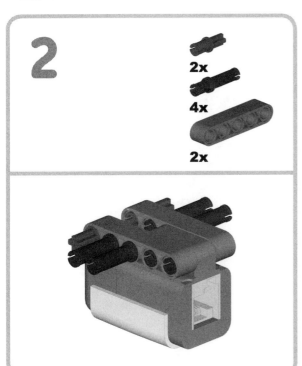

3

2

2x

2x

4

2x 3

1x

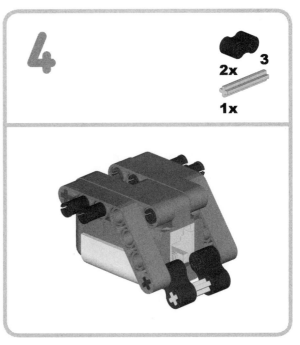

5

2x
1x

6

2x
1x

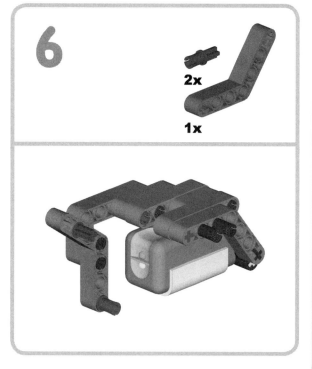

7

1x

8

2x

9

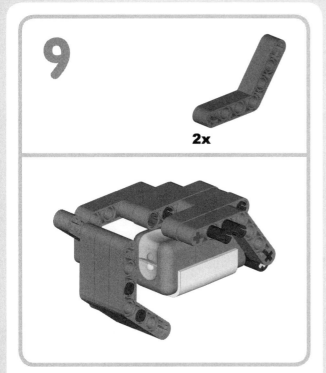

2x

10

2x

11

2x
1x

12

1x

13

1x

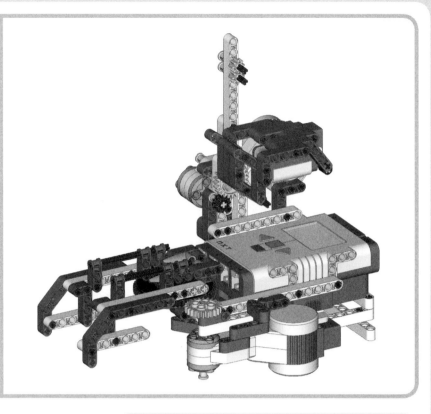

the store

1

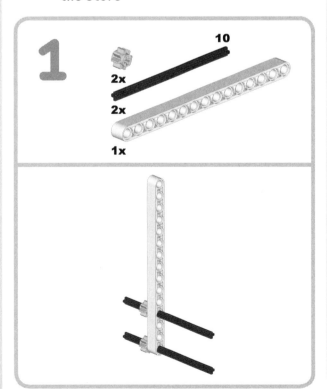

2x
2x
10
1x

2

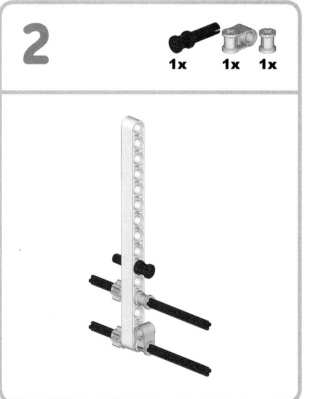

1x   **1x**   **1x**

3

1x 10 1x
1x 1x

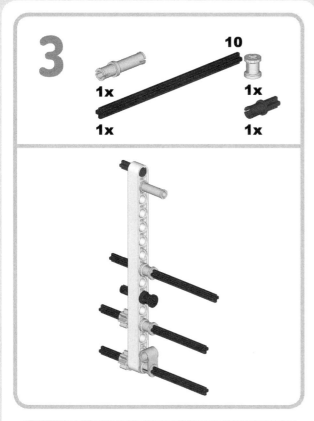

4

1x 1x
1x 2x

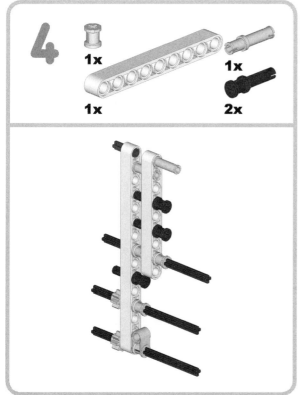

5

2x
1x 1x

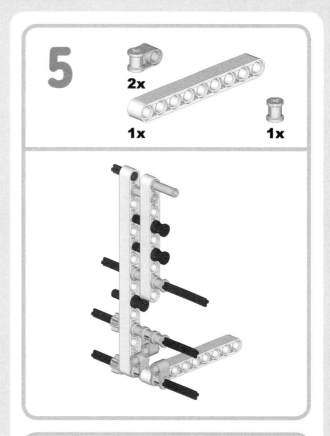

6

1x 1x
1x 1x
1x 2x

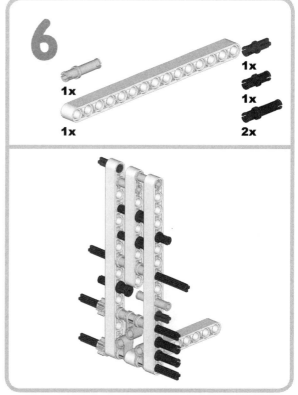

7

1x
1x

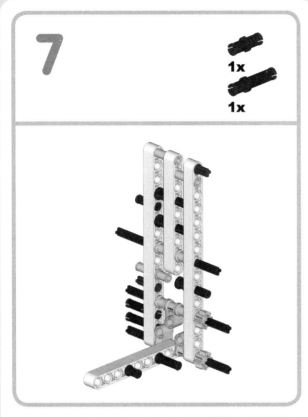

8

2x
1x

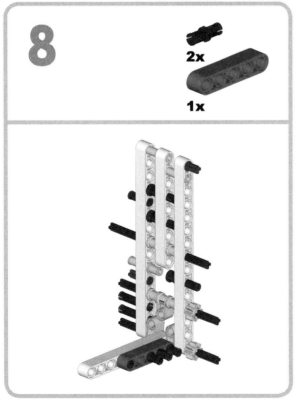

9

1x
2x

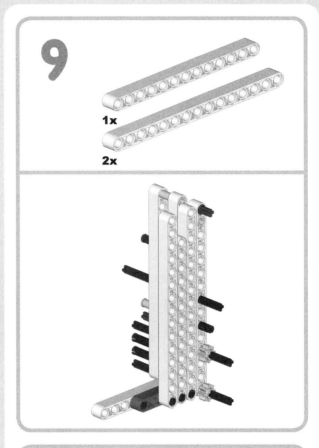

10

1x
1x
1x

10

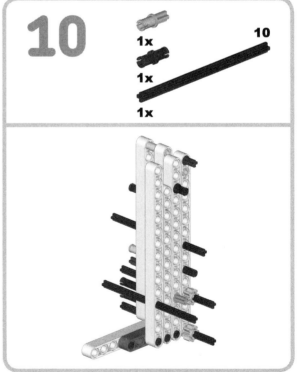

11

1x 1x

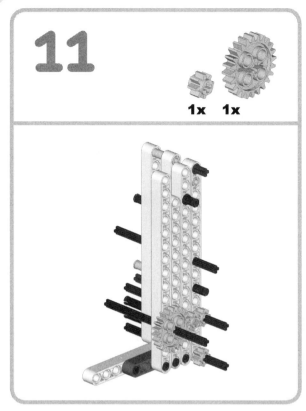

12

2x 1x 1x

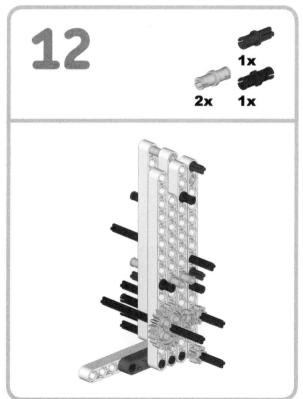

13

2x 1x 1x

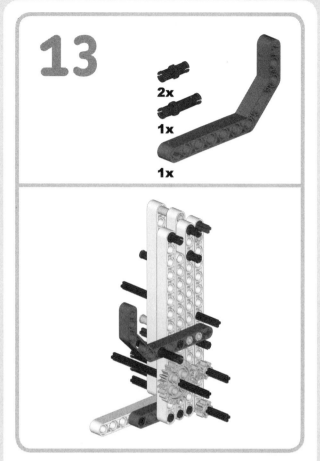

14

1x 1x 2x

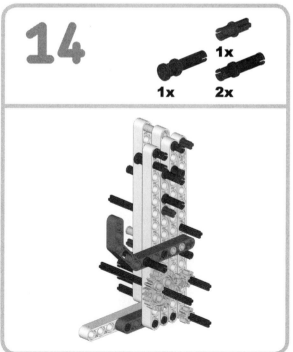

15

2x 1x 1x

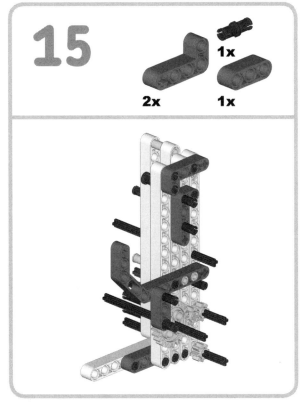

16

1x

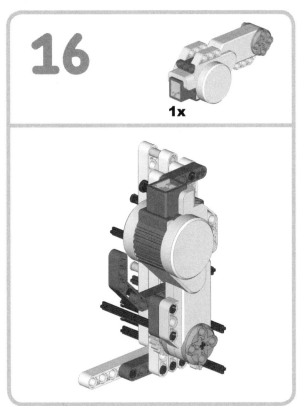

17

1x 1x 1x 5

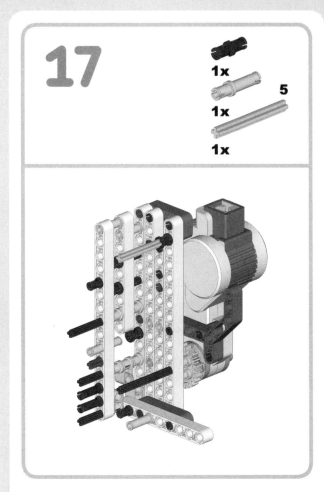

18

1x 2x

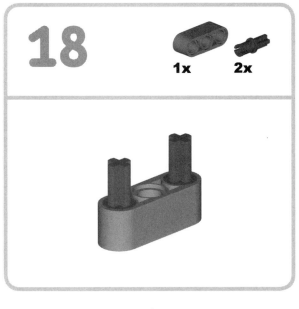

19

2x

20

1x **1x**

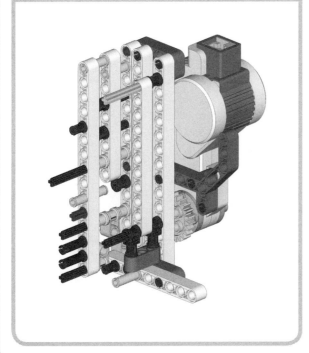

21

4x

1x

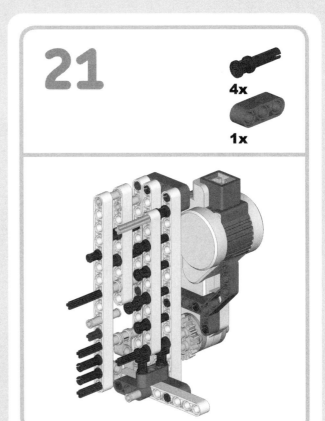

22

1x **2x** **2x**

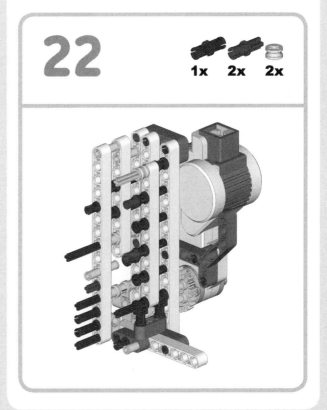

23

3x
1x
1x

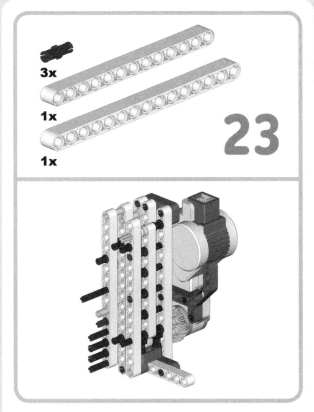

24

1x
1x
1x
2x
1x

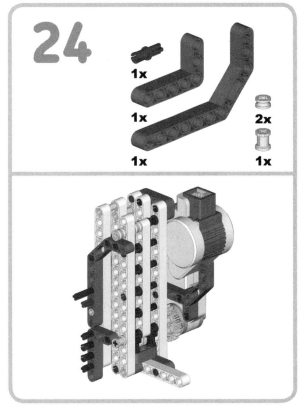

25

1x
1x

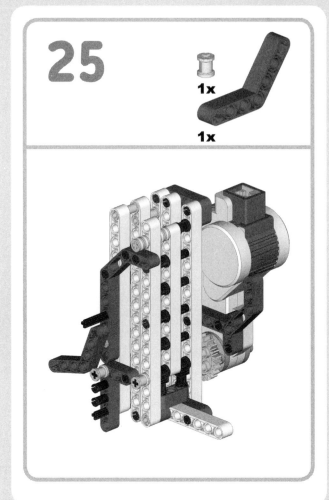

26

2x

1x

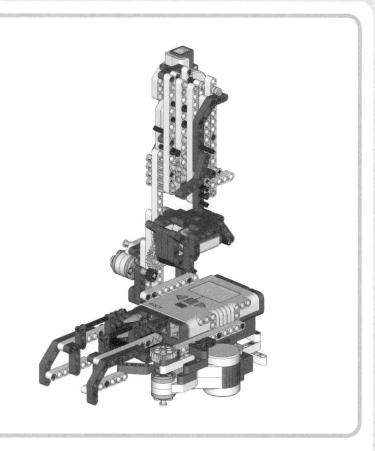

27

1x **2x**
2x

28

2x

1x

29

3

1x

1x

1x

1x

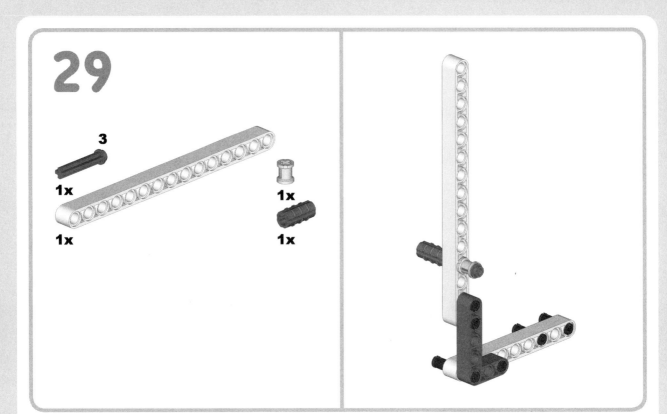

30

1x

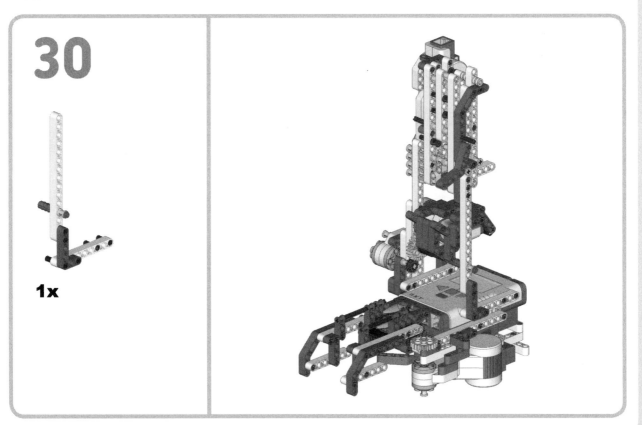

the store's cover

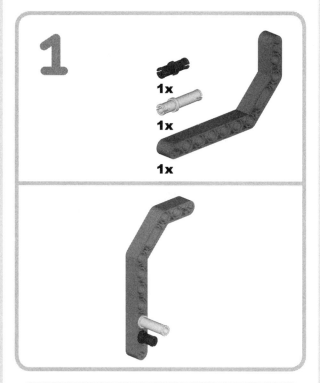

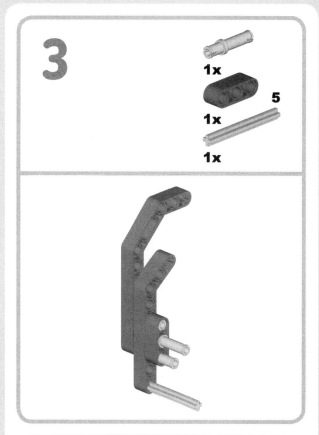

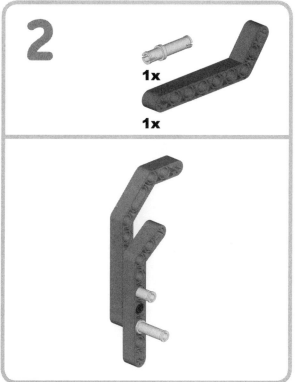

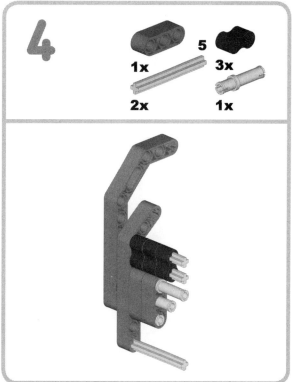

5

3x

1x

1x

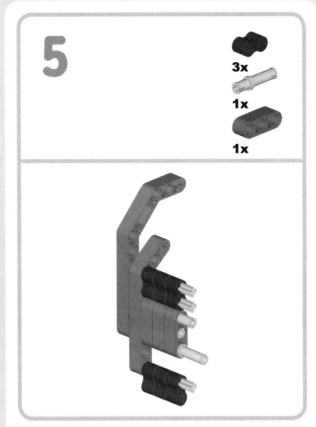

6

1x

1x

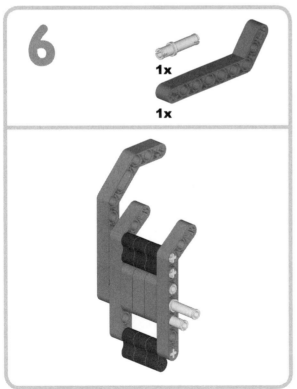

7

1x

1x

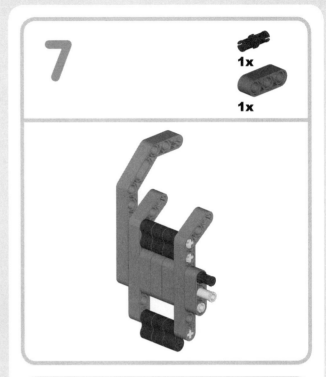

8

1x

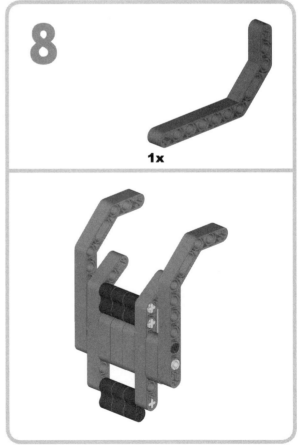

9

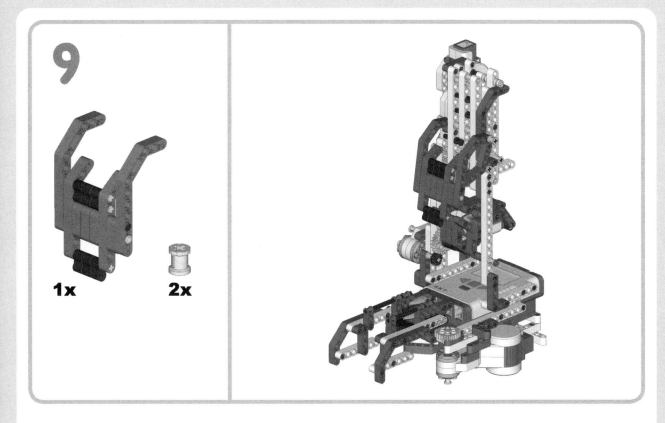

1x   **2x**

Finally, we connect the motors to output ports A, B, and C of the Brick, and we connect the Light Sensor to input port 1, as shown in Figure 3-6.

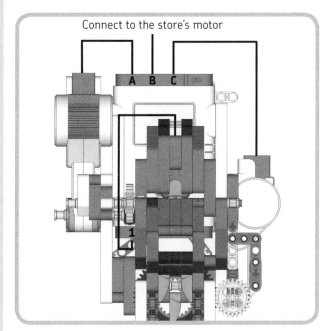

Connect to the store's motor

Figure 3-6: Connecting to the Brick

programming m

The following sections will walk you through programming your M&M sorter.

determine the reflection values for each candy color

To determine the color of an M&M, we first measure the degree of light reflected by each candy color (using the Light Sensor). Then, having determined a range for each color, we use our observed values to derive the color of the M&M based on readings of the light it reflects.

However, because the value of reflected light depends on external factors (like ambient light), we first need to measure light in the robot's environment so that we can use that as a baseline.

To do so, we use a small program that will display the value of light reflected by each candy color on the Brick's LCD.

To retrieve the values of reflected light for each candy color, we start the program we created in Figure 3-7, adjust the scanner unit (see the following section on running the main program), put an M&M into the scanner unit, and lower the scanner slightly so that the Light Sensor's receptor is directly over the candy. The amount of light reflected by the particular M&M color will be displayed on the Brick's LCD. Repeat the measurement of reflected light for each candy color multiple times, and note the upper and lower bounds of the values. We'll use the values in the main program later.

NOTE The Light Sensor will be able to distinguish colors much more reliably if you dim the light in the room in which the robot is working.

For our particular environment, we measured the following values of reflected light (the values in your environment might differ):

* Yellow: greater than 50
* Red: between 40 and 50
* Brown: between 35 and 40
* Green: between 30 and 35
* Blue: below 30

Now we turn to writing the main program to sort M&Ms.

the program for the sorter

To understand what we want our program to do, see the diagram in Figure 3-8, which shows the processing of the different actions performed by M as it sorts the M&Ms. The program will enable M to perform the following activities:

1. Drop a new M&M from the store into the scanning unit.

2. Scan for color.

3. Adjust the collector as necessary (the sorting step).

4. Drop the M&M into the sorter unit (the collector).

5. Reset the collector to its initial settings.

These activities are processed in a step-by-step (*sequential*) flow, as shown in Figure 3-8. This flow is performed in a loop until there are no more M&Ms left to sort.

With our diagram in hand, we can now decide which My Blocks we'll need to create for each step. We'll use beams and loops to connect these blocks as we create the program's flow.

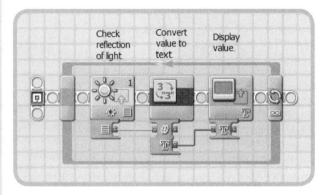

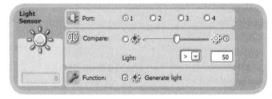

Figure 3-7: Determining the values of reflected light for each color

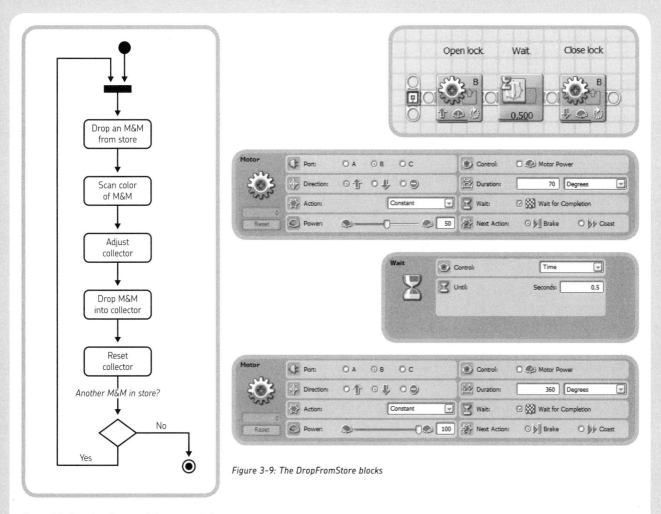

Figure 3-9: The DropFromStore blocks

Figure 3-8: Overview diagram of the program's flow

my block #1—dropping an m&m from the store

To get an M&M from the store, we run the motor that drives the store's lock as follows:

1. Drive the motor 70 degrees backward.

2. Wait a half second for the M&M to fall out.

3. Close the lock by running the motor 70 degrees forward.

Having created your My Block, name it *DropFromStore*.

my block #2—determining the color of the m&m

To determine the color of the M&M, we do the following:

1. Use the Light Sensor to compare the amount of light reflected by the M&M with each of our measured values.

2. Store the resulting color in a number variable. Each number represents a particular color.

NXT-G does not provide a *case* block, that is, a block that allows a program to choose more than two different branches of processing in response to different values of a condition. Our only option is to use the *condition* block, which can process only two different branches. Hence, programming our robot to sort between five possible colors gets somewhat complicated. The solution we arrived at is to use a *nested* sequence of condition blocks (meaning blocks within other blocks) as shown in the following programming steps.

NOTE Notice that here we use the reflection values for the colors that we retrieved with the previously described helper program. If the values in your environment differ, replace the values with your own in the following condition blocks.

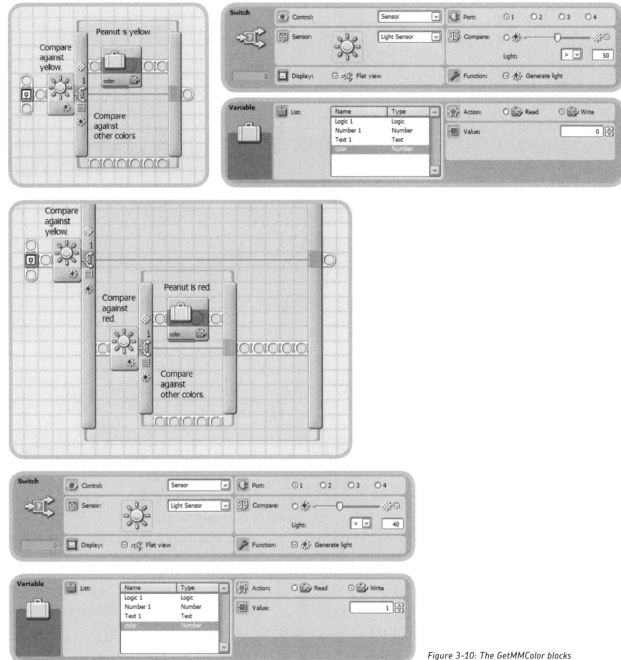

Figure 3-10: The GetMMColor blocks

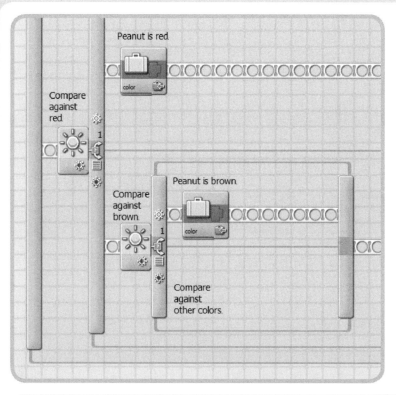

Figure 3-10 (continued): The GetMMColor blocks

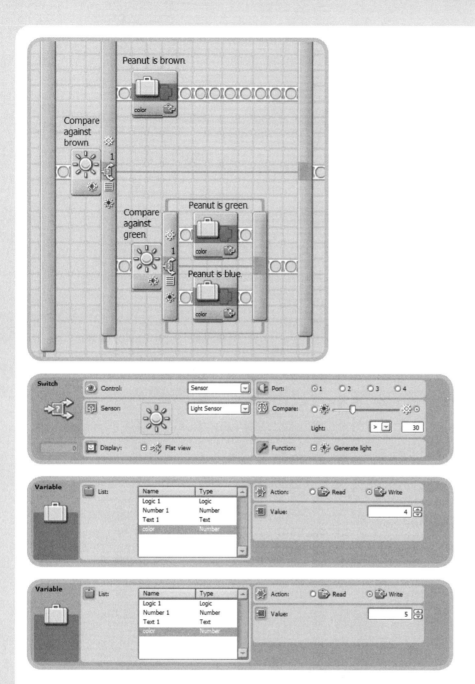

Figure 3-10 (continued): The GetMMColor blocks

Name this My Block *GetMMColor*.

my block #3—scanning the m&m

To scan the M&M that has been dropped into the scanner unit, do the following:

1. Lower the unit somewhat to receive better scanning results.

2. Read the Light Sensor to determine the color of the M&M.

3. Pause to ensure that the color has been read correctly.

4. Lift the scanner unit.

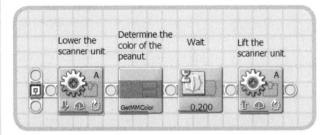

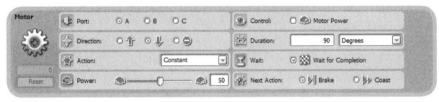

Figure 3-11: The CheckMMColor blocks

NOTE Notice that in this program we have used the GetMMColor block we created previously.

Name this My Block *CheckMMColor*.

my block #4—sorting

In the context of M, *sorting* means positioning the collector at the front of the robot and adjusting it to ensure that each piece of candy will follow the correct path when it is released from the scanning unit.

Determining the correct position of the sorter depends on the color-code number we have stored in the variable set with a value in My Block #2; hence, the program simply compares a recorded measurement of the light reflected from an M&M against those baseline values. Again, we use a nested sequence of condition blocks.

From the comparison, we deduce the degree and the direction we have to run the collector's motor to position the collector appropriately.

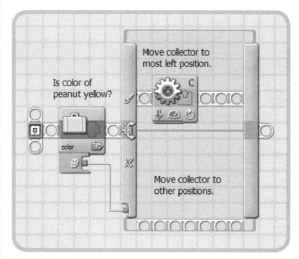

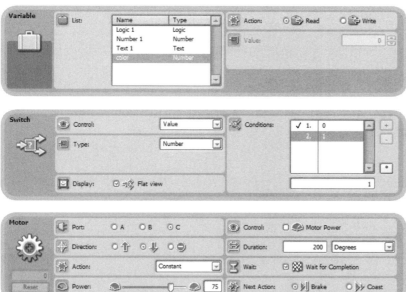

Figure 3-12: The Sort blocks

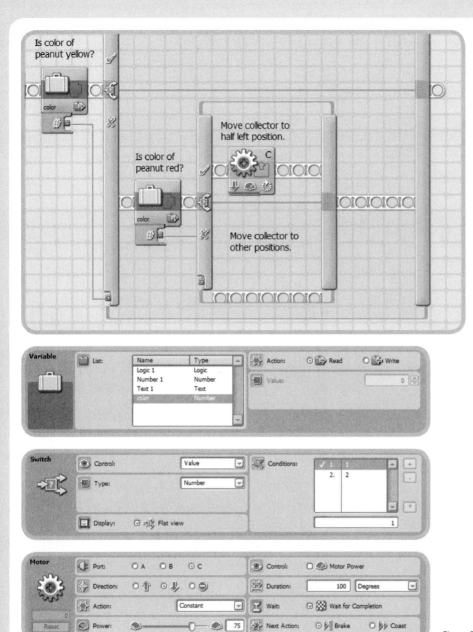

Figure 3-12 (continued): The Sort blocks

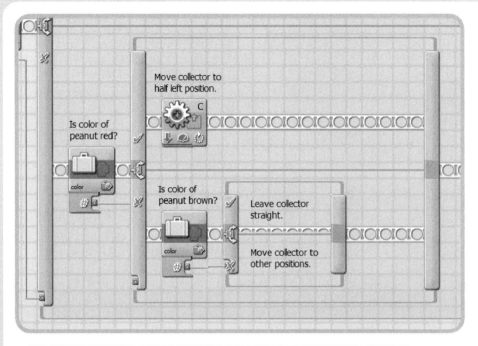

Figure 3-12 (continued): The Sort blocks

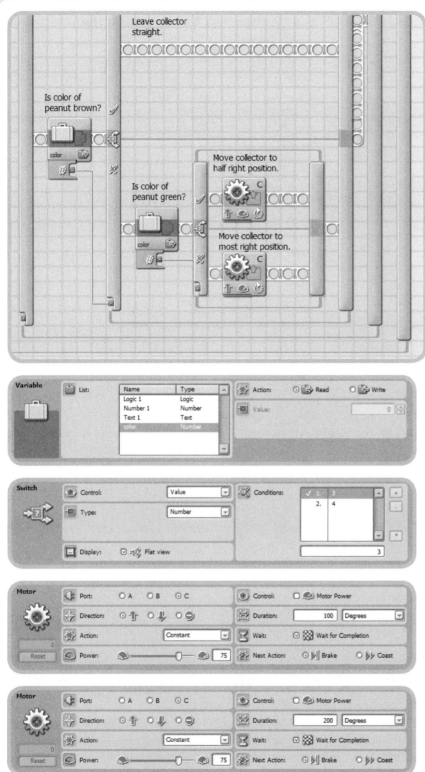

Figure 3-12 (continued): The Sort blocks

Name this My Block *Sort*.

my block #5—releasing the m&m

After adjusting the collector, we want to release the M&M from the scanning unit and drop it into the correct position, led by the collector. To achieve this, we:

1. Drop the scanner unit by turning the connected motor 150 degrees.

2. Pause so the M&M can drop out.

3. Reset the scanner unit to its initial state by running the motor backward 150 degrees.

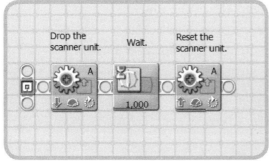

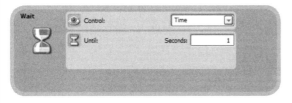

Figure 3-13: The Release blocks

Name this My Block *Release*.

my block #6—resetting the connector

Now the only thing left to do is to reset the collector to its initial state. To do this, we simply reverse the sorting step by copying the contents of the Sort My Block and changing the motor direction in each Motor block.

Name this My Block *ResetCollector*.

main program

We now have everything we need to assemble the main program. Basically, the program consists of two elements:

1. A loop that runs as long as there are still M&Ms in the store

2. A sequential process to sort one M&M

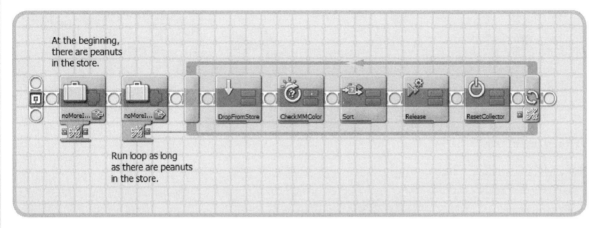

At the beginning,
there are peanuts
in the store.

Run loop as long
as there are peanuts
in the store.

Figure 3-14: This is the main program of M. Because this version of the program doesn't actually check to see if there's still an M&M in the store, the program will run until you manually stop it. We leave it to you to modify the program to solve this challenge. (You'll find some possible solutions in "Further Exploration" on page 95.)

running the program

Before we run the program, reset M to its initial state as follows:

1. Close the store's lock.

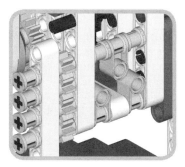

Figure 3-15: Initial state of the store's lock

2. Position the scanning unit parallel to the floor so it is ready to take an M&M from the store.

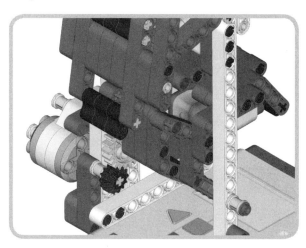

Figure 3-16: Initial state of the scanning unit

3. Move the collector to its middle position, pointing straight ahead.

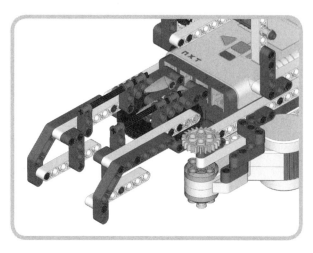

Figure 3-17: Initial state of the collector

4. Fill the store with M&Ms.

Now we can start the program—happy sorting!

further exploration

Now that M is working and you have all your M&Ms sorted (and eaten), you might consider some further enhancements to the robot.

check for an empty store

As already discussed, M's program doesn't actually check to see if there's still an M&M left in the store; in other words, the variable noMoreInput is never set to True. As a consequence, the program runs until you manually switch it off, even if there are no more M&Ms left to be sorted.

To overcome this, M would have to detect that the store is empty, and you would need to set the variable noMoreInput to True in such a case. One way to achieve that is to mount an additional Light Sensor on the store to scan for reflections of candies at the beginning of each run of the loop. Another option might be to use the scanner's existing Light Sensor. If the measured value of reflected light is very low, M could deduce that there's no candy in the scanner unit at all and that the store must be empty.

use a color sensor

You may notice that the Light Sensor is not very reliable when it comes to color detection. Because the values of reflected light are affected by ambient light, M is prone to misread colors.

The best solution for this problem is to use a special Color Sensor, available from companies like HiTechnic (*http://www.hitechnic.com/*). HiTechnic can also provide you with the NXT-G programming block you will need to use the sensor.

When replacing the Light Sensor with a Color Sensor, you will need to modify your program accordingly. When you begin using it, you will find that M is much more reliable.

enhance the collector unit

To build M with a single NXT retail kit, the collector unit has to be kept very simple so that it simply sorts the candy. The M&Ms themselves are not collected in separate color-related containers. However, if you have additional TECHNIC and NXT parts, consider providing a container for each color. Each of these containers could be programmed to position itself below the scanning unit in the sorting step, using a circular collector or a horizontal rail with different containers attached.

enhance the store

Due to the different sizes and sometimes irregular shapes of the M&Ms, the store in its present design is prone to jam when an M&M gets stuck in the lock.

You might consider alternative designs of a store that are less susceptible to jamming. After all, your choice of different designs or of further enhancements to M are limited only by your imagination!

NXT dragster: the NXT STEP dragster challenge

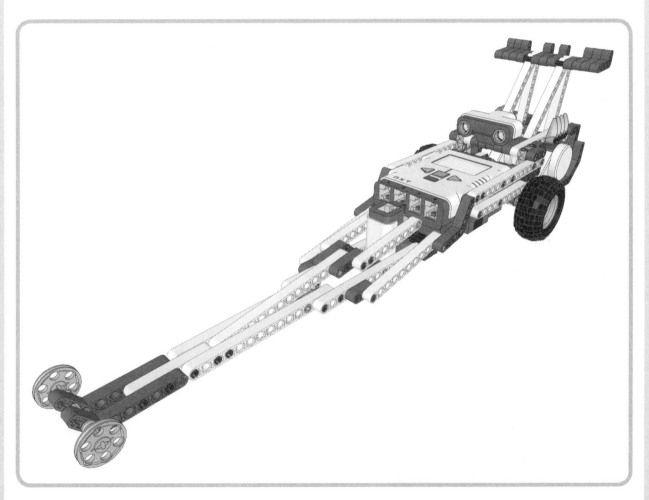

Figure 4-1: The NXT Dragster

Welcome, race fans! Gear up for the drag strip as you create the NXT Dragster and enter into the newest NXTreme LEGO MINDSTORMS motorsport . . . The NXT STEP Dragster Challenge.

You're joining the one-kit race series where the NXT is the king of the drag strip!

To reach the winner's circle, you must first create a winner. In this chapter, you'll find the essentials that will prepare you for a day at the drag races, and it will require only the LEGO elements from your NXT kit.

We'll also show you how to recreate the official drag strip layout, which will be used to measure your vehicle's performance as you power down the drag strip. You will record the elapsed time of each drag run for a place in the online Starter Series Records Vault, and as your need for speed grows, "Going Faster Starts in the Pits" on page 131

will fuel your drive to enhance your vehicle's performance and push the limits.

Success at the finish line is a culmination of skilled engineering and execution by the technical designer, but there are limits to what can be achieved with one NXT kit. Reaching those limits and conquering them will be the challenge. Are you ready? This chapter will teach you the skills you'll need to succeed.

building the NXT dragster

The experience you gain from building and racing the NXT Dragster will spark a desire to develop your own winning

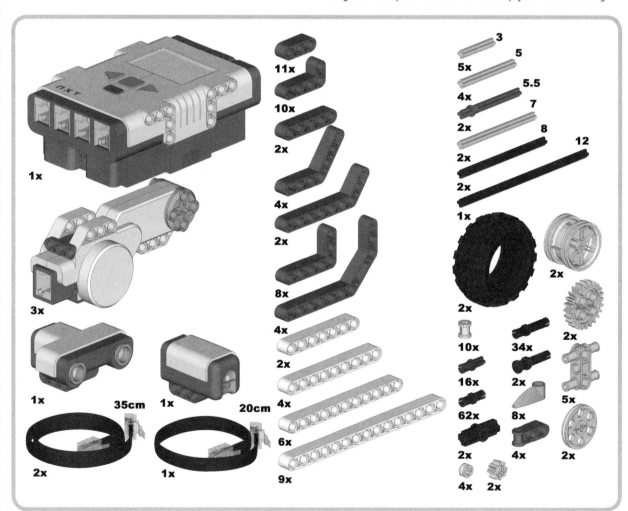

Figure 4-2: Bill of Materials

design. This is one point in the process where your imagination is very handy, and I encourage you to begin working on your own design after running the original NXT Dragster down the drag strip a few times. Now, grab your kit and let's build!

The NXT Dragster is constructed from one NXT kit using the list of parts in Figure 4-2, so gather the necessary parts and follow these building instructions carefully. You'll find tips throughout the chapter to aid you in the project.

To trigger the run timer, the NXT Dragster uses an onboard downward-facing Light Sensor to detect the starting line and finish line of the drag strip. The Light Sensor can be positioned anywhere onboard the vehicle as long as the light reading area is not more than three-fourths of an inch above the drag strip surface.

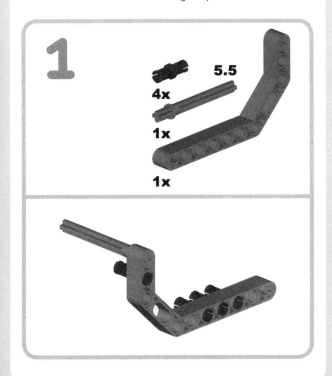

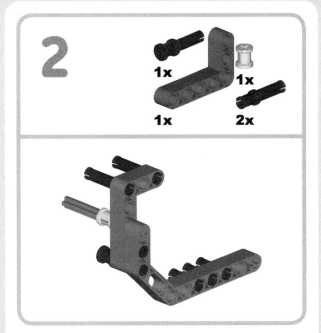

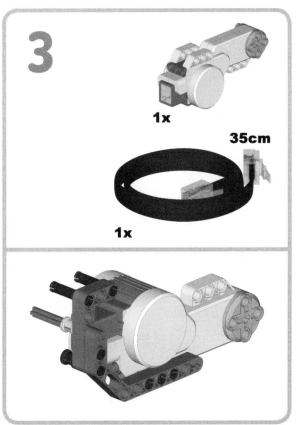

The cable shown in step 3 should be connected to this motor and will later be connected to output port B of the NXT Brick.

4

1x
2x

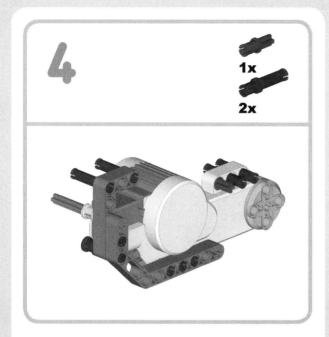

5

2x
1x

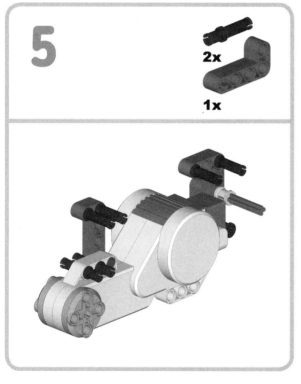

6

3x
5
1x
1x

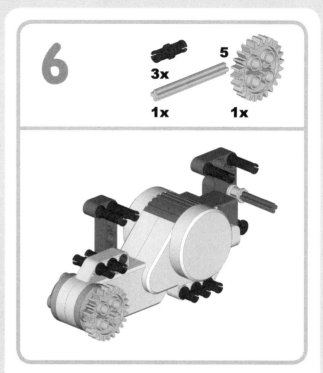

7

1x
2x

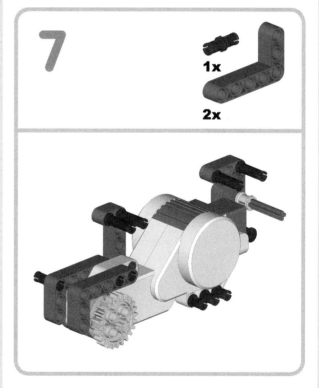

8

1x
8
1x
3x

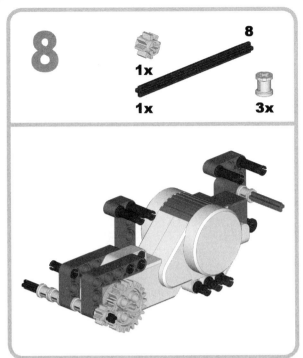

9

3x
1x

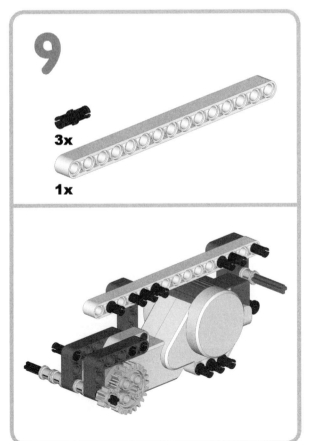

10

7
1x

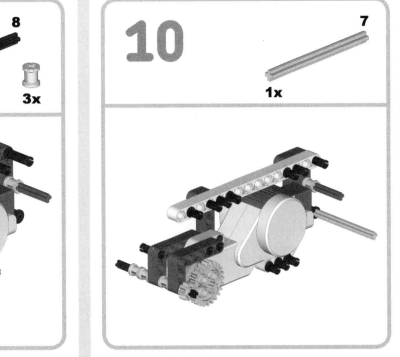

11

1x
1x

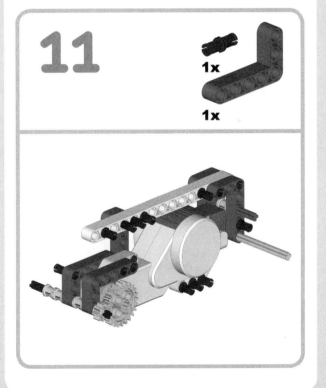

12

3

1x

2x

1x

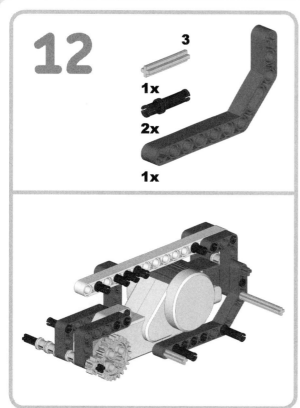

14

4x

1x

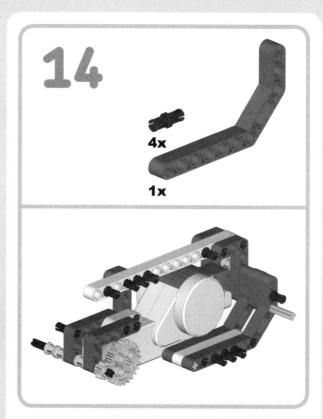

13

1x

1x

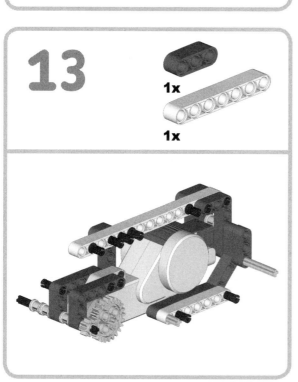

15

2x

1x

1x

16

1x
1x
1x

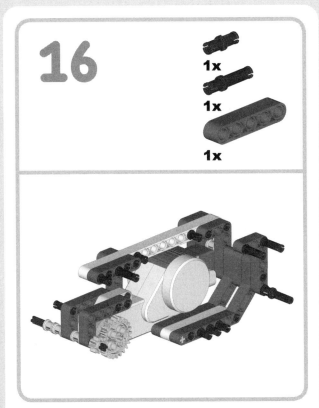

18

1x
1x

NOTE The Ultrasonic Sensor does not use a cable in this design.

17

1x
1x

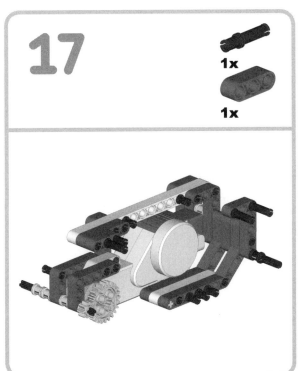

19

1x
1x

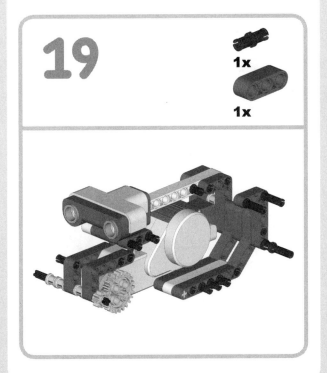

20

1x

2x

1x

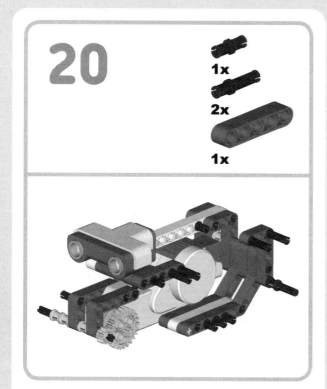

21

2x

1x

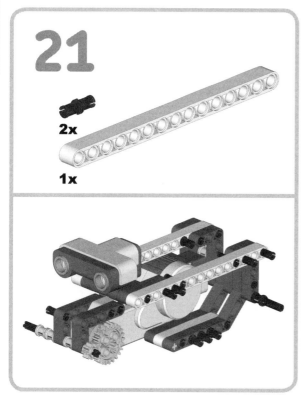

22

1x

1x

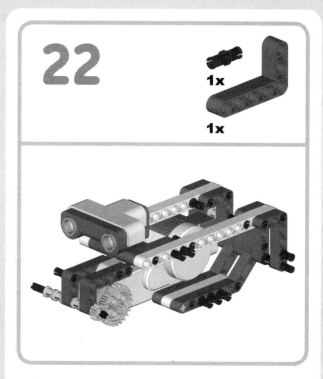

23

5.5

1x

1x

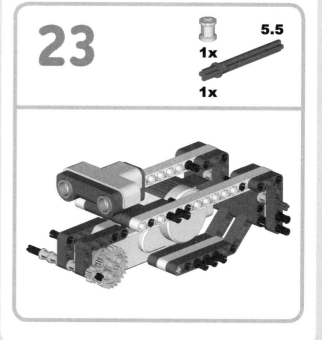

24

1x

35cm

1x

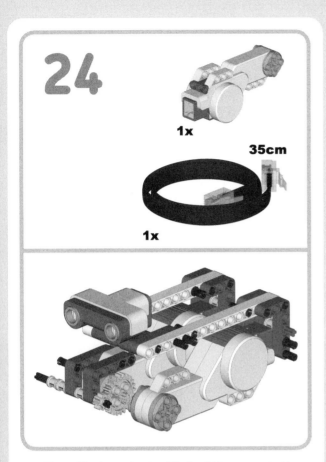

The cable shown in step 24 should be connected to this motor and will be connected later to output port C of the NXT Brick.

25

5

1x   1x

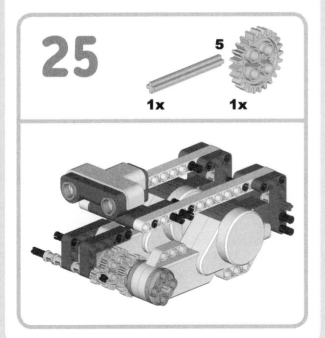

26

2x   1x

1x   3x

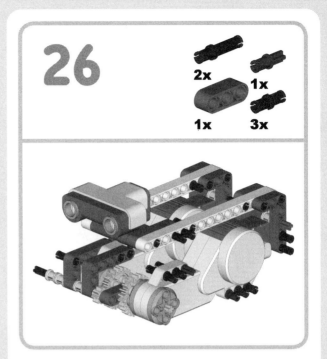

The 3-hole beam in step 26 should hold the ends of the three axles and the axle added later in step 28.

27

1x

1x

2x

1x

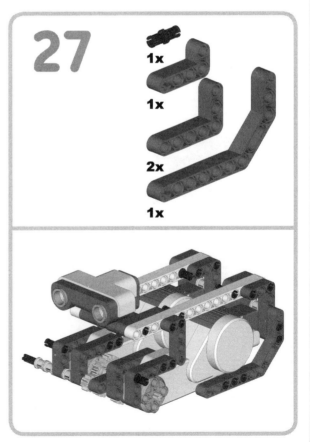

28

8

1x

1x

3x

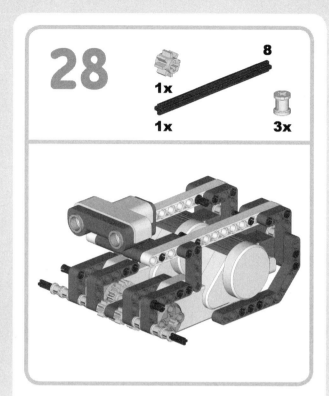

29

2x

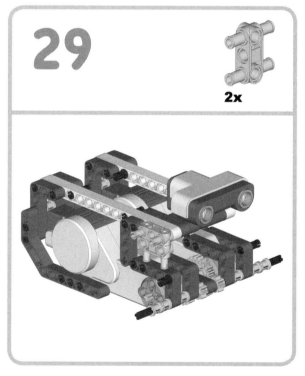

30

2x

1x

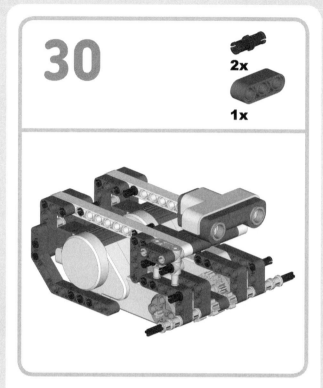

6x

1x

1x

31

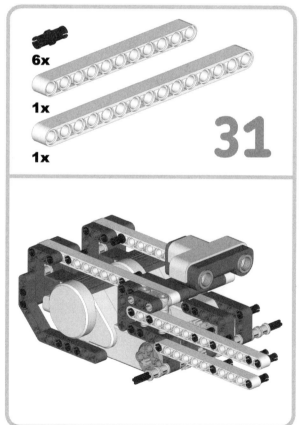

32

1x

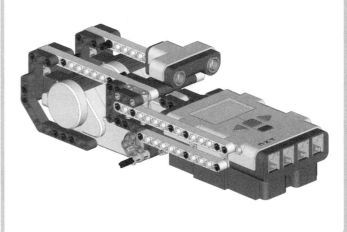

33

2x

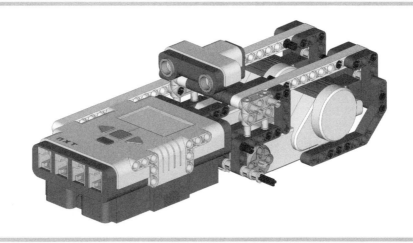

34

2x

1x

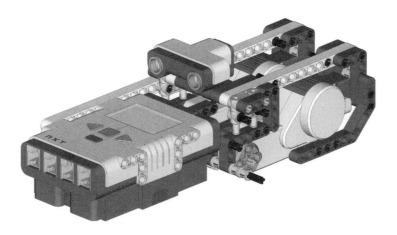

35

6x

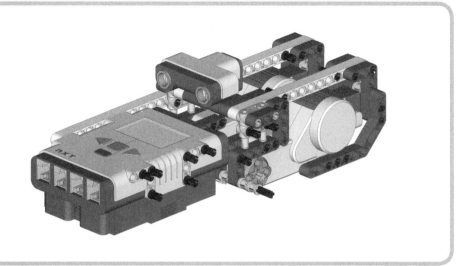

36

1x

1x

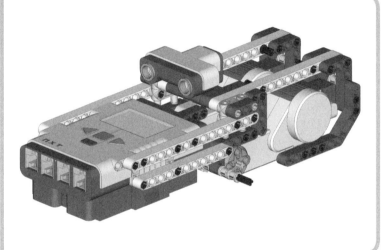

37

2x

1x

38

1x

1x

39

2x

1x

40

1x

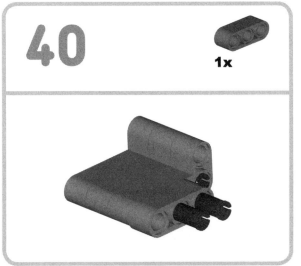

41

1x

12

1x

1x

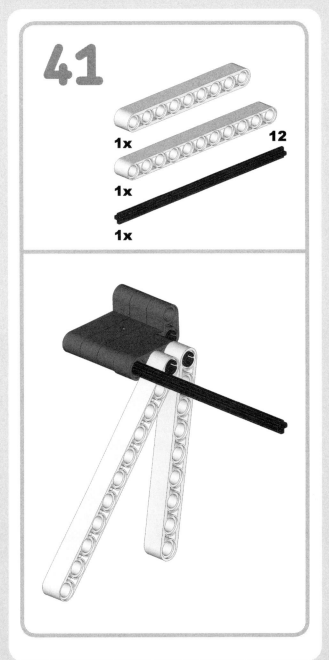

42

1x
1x

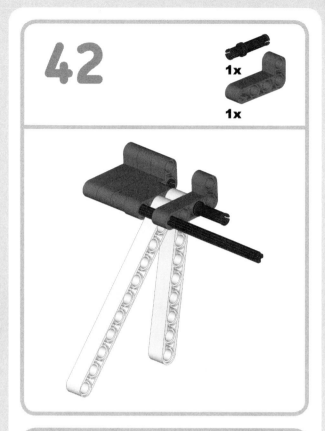

43

1x
1x

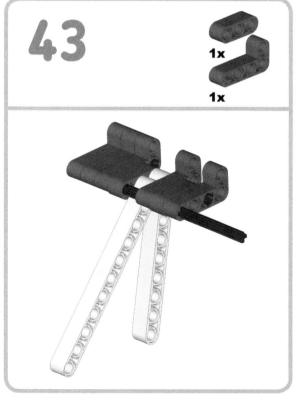

44

1x
1x
2x
1x

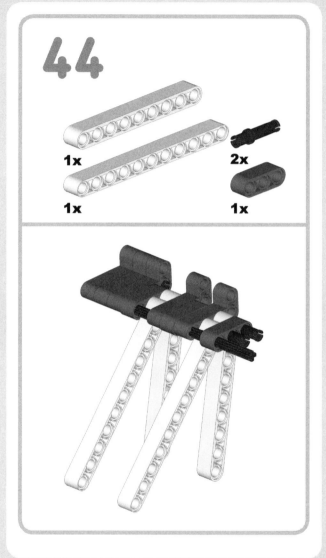

45

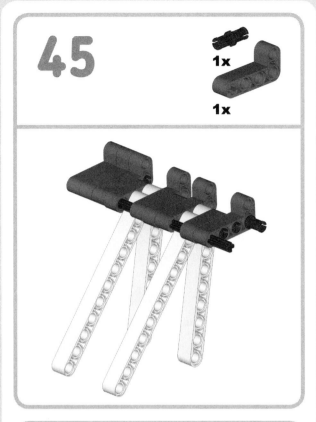

46

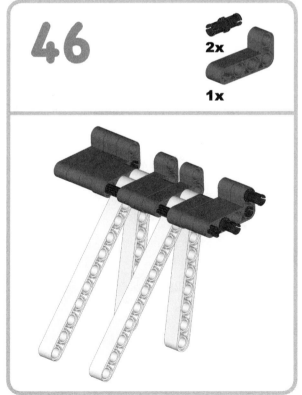

47

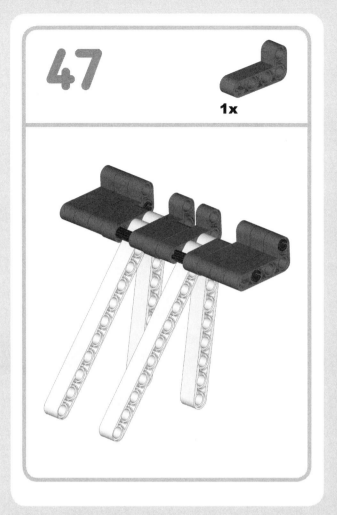

48

1x

49

8x

8x

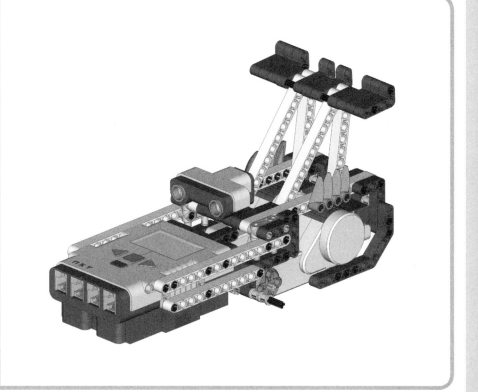

50

4x

3x

1x

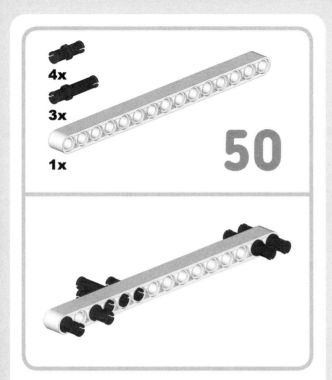

52

3x

1x

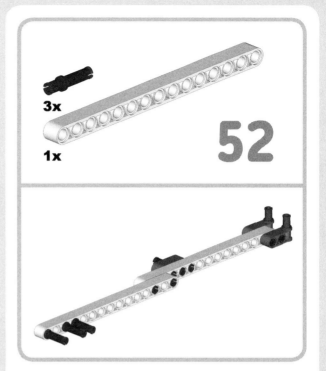

51

2x

2x

1x

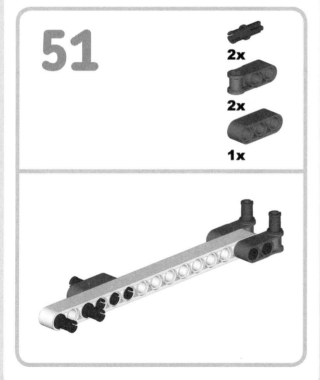

53

2x

1x   1x

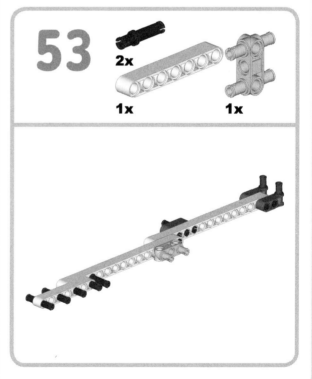

54

2x

55

1x **5**

1x

1x

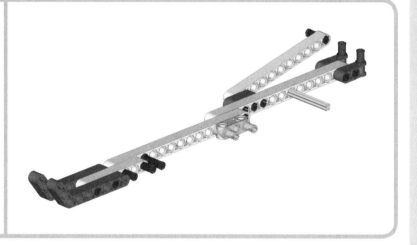

56

2x **3**

1x

1x

1x

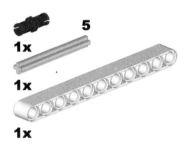

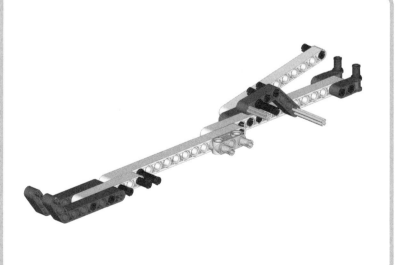

57

1x

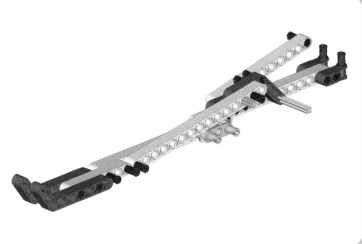

58

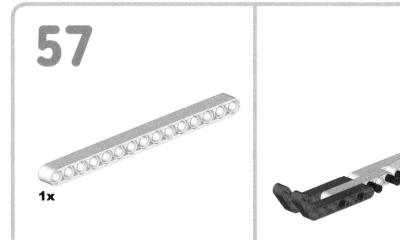

2x

1x

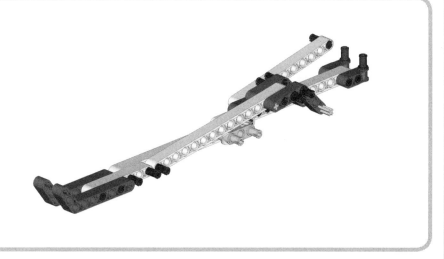

59

3

2x

1x

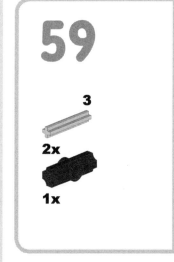

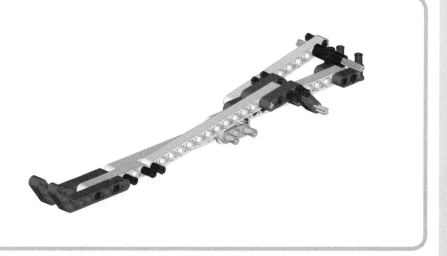

60

1x

5 **1x**

1x

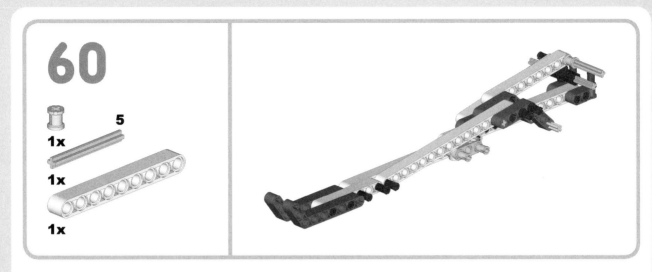

61

20cm

1x

1x **1x**

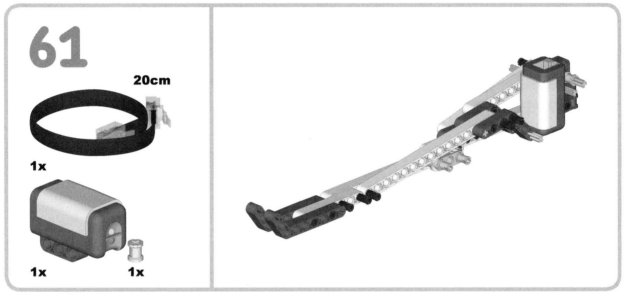

The cable in step 61 is connected to the Light Sensor and will later be connected to input port 3 of the NXT Brick.

62

3

1x

1x

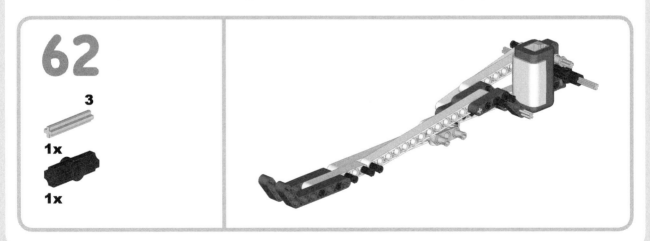

63

1x

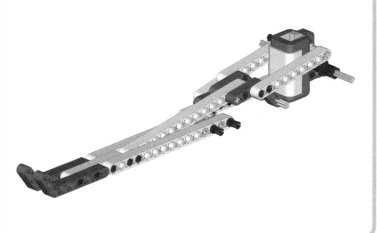

64

2x

1x

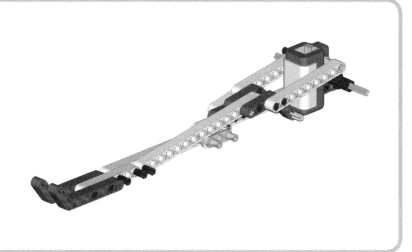

65

2x

1x

1x

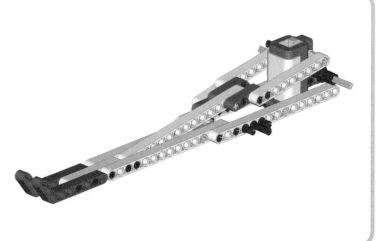

66

 1x

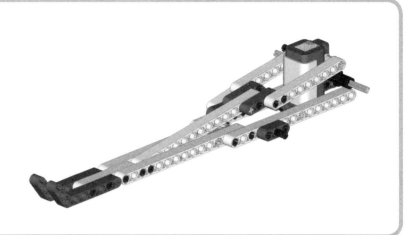

67

 1x

1x

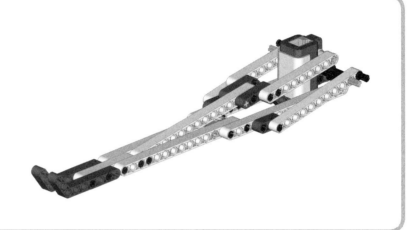

68

 2x

69

2x

2x

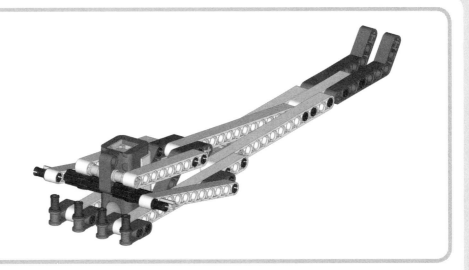

70

1x

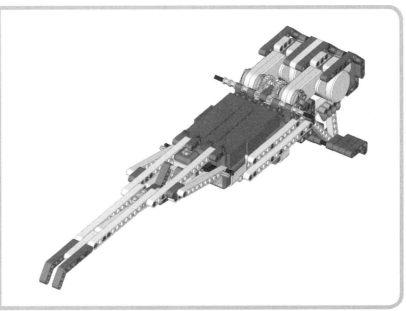

71

1x

1x

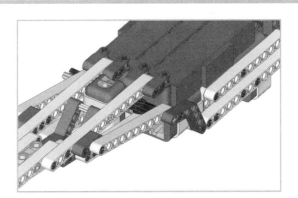

72

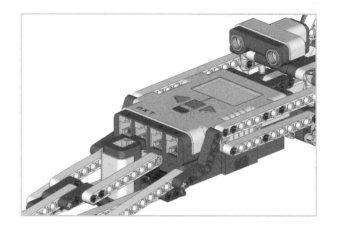

1x

1x

73

2x

2x

2x

2x

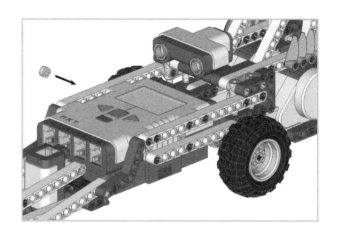

74

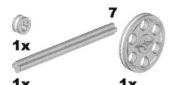

1x

7

1x

1x

75

1x

76

1x   1x

Figure 4-3: The completed NXT Dragster

program-ming the NXT dragster

Now that the NXT Dragster is assembled, let's create the program for a speedy trip down the drag strip. The *SSDragRun.rbt* NXT-G program is a fairly simple and straightforward automatic sequence that controls the NXT Dragster throughout one drag run on the standardized drag strip.

This program positions the NXT Dragster at the starting line, counts down to the start, and records the run time until the NXT Dragster crosses the finish line. The elapsed time is then displayed until you stop the program. All of the programming blocks should be arranged on the same sequence beam, as illustrated in Figure 4-4. Follow the images closely, and pay attention to the notes included with the instructions because they will help you understand the program.

Go ahead and play with the program—just remember that the original standardized program code should be used when recording times for entry into the online Starter Series Records Vault.

The Light Sensor block illustrated in Figure 4-4 generates light; it also samples the reflective conditions of the drag strip surface to establish the current lightness reading when the NXT Dragster is in the clear staging area behind the start line. This reading is the limit or threshold used to trigger when the NXT Dragster is to detect a black line (or a line that is darker than the lightness reading). The Light Sensor is connected to input port 3 on the NXT Brick, but you can choose to use a different port to fit your design and still qualify for the Starter Series—just be sure to make that change in your program.

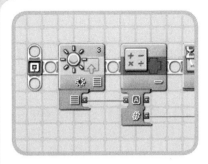

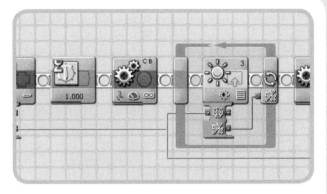

Figure 4-4: The first set of programming blocks calibrate the NXT Dragster to the drag strip surface.

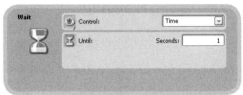

Figure 4-5: This group of blocks allows the NXT Dragster to find the starting line.

The Math block illustrated in Figure 4-4 receives the light reading, subtracts a little to create a darker trigger point, and pipes this trigger point value to the Light Sensor blocks downstream via data wires. This calibration is performed every time the program is started.

The programming blocks illustrated in Figure 4-5 command the NXT Dragster to find the starting line using the established light calibration. A short pause from the Wait block will allow you to clear your hands away after starting the program. The NXT Dragster then slowly rolls forward until the Light Sensor detects the black starting line, which breaks the loop. The Move block controls motors B and C and can be altered to match the motor arrangement of your design, which still qualifies it for the Starter Series.

The programming blocks illustrated in Figure 4-6 command the NXT Dragster to stop the motors when it detects the starting line. The prestart countdown of three beeps in a loop begins at this point.

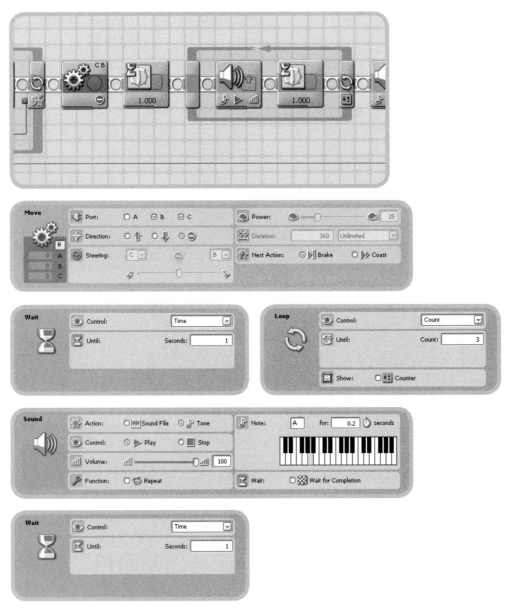

Figure 4-6: This group of blocks stops the NXT Dragster at the starting line and begins the countdown.

The programming blocks illustrated in Figure 4-7 complete the countdown with one long and high beep, simulating the "Go!" signal. The NXT Dragster heads down the drag strip with full motor power, and the timer begins recording the elapsed time of the run. The Wait block in Figure 4-7 allows the Light Sensor to pass the starting line before it begins to look for the finish line.

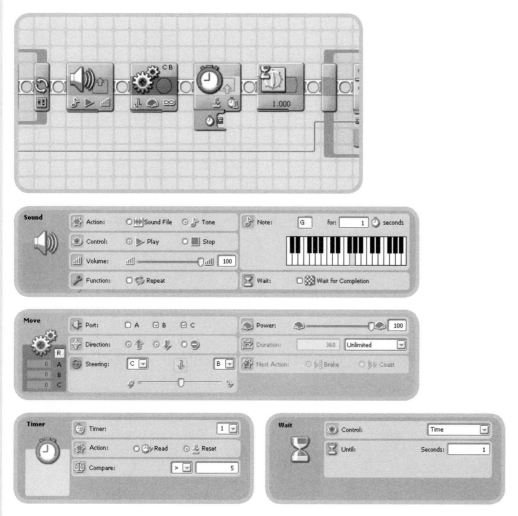

Figure 4-7: This group of blocks signals the start and begins timing the run.

The group of programming blocks illustrated in Figure 4-8 commands the Light Sensor to look for the finish line and stop the motors (allowing them to coast a bit) when it detects that the NXT Dragster has crossed a black line. Crossing a boundary line early will have the same result.

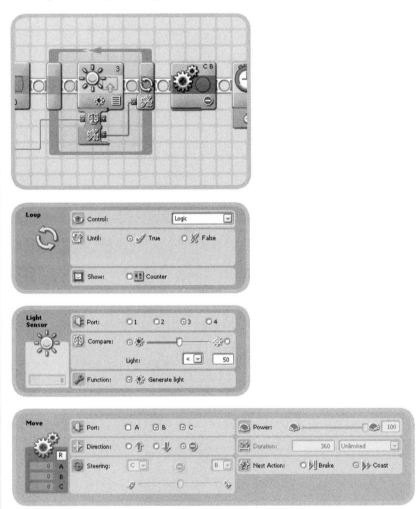

Figure 4-8: This group of blocks controls the NXT Dragster when it's passing the finish line.

The final group of programming blocks illustrated in Figure 4-9 stops the run timer and displays the elapsed time of the drag run in milliseconds on the NXT screen. The program will continue to display the time for you to record until you press the Enter button on the NXT.

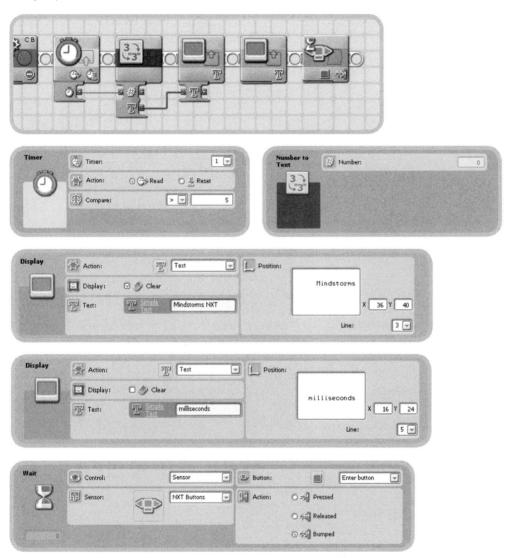

Figure 4-9: This group of blocks stops the countdown and displays the elapsed time on the screen.

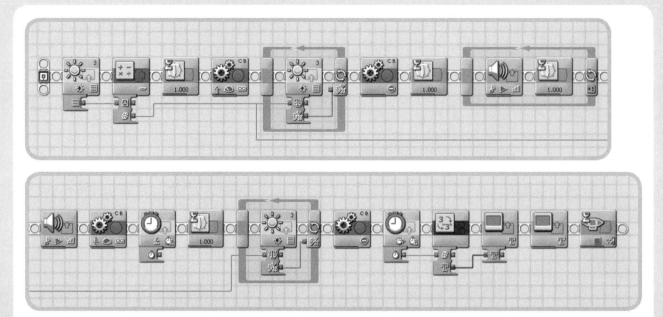

Figure 4-10: The complete NXT Dragster SSDragRun NXT-G program

This completes the Starter Series programming sequence. Be mindful of the control panel settings for each programming block, as well as the data wire connections. Everything must be arranged exactly as shown in the illustrations for the NXT Dragster to function properly. Save it to your computer and compile it to your NXT Brick.

making the drag strip

When you've created the *SSDragRun.rbt* NXT-G program, you'll need a drag strip to race down, so let's make it. The drag strip is the domain used to measure and track the NXT Dragster's performance. Figure 4-11 details the standardized 20-feet-long-by-3-feet-wide clear area for the drag strip layout and boundary measurements. To construct this layout you must first find a suitable space—a fairly smooth, seamless, level surface that is a light color and free of obstacles, dark spots, and patches. It should measure approximately 26 feet long by 8 feet wide (add 2–3 feet to the width for side-by-side drag strips). The best choice indoors is a hallway with a white floor; the best choice outdoors is a light gray concrete driveway. Create the boundaries with common three-fourths-inch wide black electrician's tape or another type of black tape, printed paper,

or even painted lines. You can increase the total black area of the finish line if you want to. *Pay close attention to the details in Figure 4-11.*

When you choose a space in which to race, it's a good idea to first test it with the following process to ensure that the NXT Dragster operates properly on the surface. Mark the space with two 12-inch strips of black tape (a starting line and a finish line), about 3 feet apart, creating a very short drag strip. Make a few drag runs. If the surface causes too many problems, find another place to test the NXT Dragster.

staging a drag run

The accelerating NXT Dragster is a wild vehicle that fights a multitude of forces to maintain a straight course and a clean finish. Getting down the drag strip as fast as possible and crossing the finish line without being disqualified by veering into a boundary marking are your main objectives. It is important to stage your vehicle properly before a run. *Staging* the NXT Dragster means placing it in the proper position on the drag strip to ensure the cleanest possible run. To help achieve your goal of a clean run, perform the following trick when you are ready to place the NXT Dragster in the staging area and before you run the program. Power up the NXT and

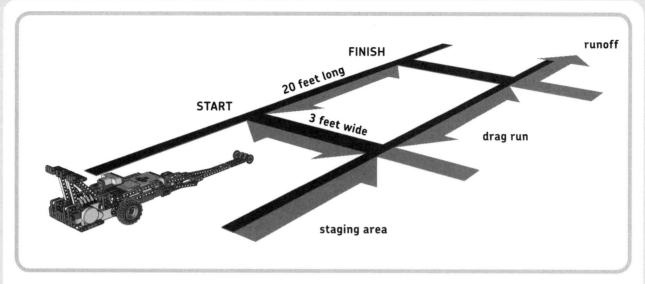

Figure 4-11: The drag strip layout

place the NXT Dragster in the staging area well behind the starting line. Pull the NXT Dragster backward, allowing the motors and gearing to engage, and remove any slack to the ground that would otherwise cause your NXT Dragster to jerk to one side. Continue rolling back a short distance (about 6 inches). This should be performed while sensing the straightest line toward the finish line. Stop, run the program, and step back. Check the heading while the NXT Dragster rolls forward to find the start line. This process gives the motors more time to synchronize and gives you a little control over how straight your line down the track will be. Practicing this technique will ensure a greater number of successful drag runs without being disqualified.

You can also improve your line by matching motors by their performance. If there is excessive resistance internal to one motor or on one side of the NXT Dragster, you'll need to work out these bugs—it's all part of drag racing. Some solutions will be explored in "Going Faster Starts in the Pits" on page 131.

following a typical drag run

The flow of a typical drag run is simple, and little is required from the driver. The programming automates the NXT Dragster's actions during a drag run, but you still need to

have a hand in the race. It is important to be familiar with the process and understand what should happen. The steps are as follows:

1. The driver powers up the NXT and places the NXT Dragster in the staging area behind the starting line (as shown in Figure 4-11).

2. The driver performs the staging maneuvers, runs the program, and steps back.

3. The NXT Dragster automatically rolls forward until the Light Sensor detects the start line.

4. The NXT Dragster pauses at the start line for the audible countdown to launch.

5. The drag run starts the run timer and continues until the finish line is detected.

6. Crossing the finish line stops the run timer and displays the elapsed time of the run in milliseconds.

7. When the NXT Dragster coasts to a stop, record the displayed elapsed time of the run before stopping the program.

If the NXT Dragster behaves differently during any portion of this process, you must restart the process. For example, a foul start could occur before or just after the start line. A vehicle that crosses a boundary marking or does not properly detect the finish line is disqualified, and you must restart the process. Successful drag runs don't come easy!

measuring time

A drag race is a contest of acceleration from a standing start, usually between two vehicles over a measured distance from start to finish. In The NXT STEP Dragster Challenge, the total length of the measured distance is 20 feet, and the time it takes for the Dragster to travel this distance is the elapsed time (ET), which will be displayed on the NXT screen after each trip down the strip. Recording and comparing the ET with other successful runs will provide an indication of how the NXT Dragster is performing.

ET is one measurement that can track performance, and it can easily be obtained with one NXT kit. As a dragster barrels down the drag strip, it constantly gains speed until it reaches the *top speed* of the run and the driver lets off after it crosses the finish line. To capture and measure the top speed, real drag strips have a *speed trap* area in the last 66 feet before the finish line, where the vehicle's top speed usually occurs. The dragster enters this area and trips a sensor that measures time, which stops when the dragster trips the finish line sensor. The top speed is quite different from ET, and the top speed is a bit difficult to obtain with one NXT kit. Capturing the top speed requires several tailored parameters to factor in the differences in wheel size and gear train designs. Can you design your own speed trap area and measure a dragster's top speed?

The *average speed* is the rate at which the dragster covers the distance where the speed may have varied. The average speed is the distance traveled by the dragster divided by the time it took to travel that distance. It is possible to calculate the average speed, but first we will convert our elapsed time from milliseconds to seconds:

$$1,000.0 \text{ milliseconds} = 1.0 \text{ seconds}$$

For example, if the dragster traveled 20 feet in 5,250 milliseconds, your converted elapsed time would be 5.250 seconds. Now, to calculate the average speed, we would use this formula:

$$\text{Average Speed} = \text{Distance} \div \text{Time}$$

So if you travel 20 feet in 5.250 seconds, your average speed of that drag run would be 3.81 feet per second (20 ÷ 5.250 = 3.81).

recording time

Time is the mark of an NXT Dragster's performance on the Starter Series drag strip, and a serious racer should keep a log book detailing the history of each successful drag run. A log book is more than just a record-keeping device full of organized data—it's a tool to let you know how each change affects your NXT Dragster and if the change is producing a gain or a loss. Any sudden new trend will be visible and identifiable in your records. If you made design changes, the best method to prove modifications is to simply make one change at a time, then test that change with a few drag runs, record the elapsed times, and compare them to previous run times. If your trend analysis hints at a drop in performance, reverse the change and try something different; repeat the process, and keep the changes that lead to improvements. The goal is to find the combinations that produce the best results.

ET is the most important recordable data point you should track in your log book. However, several design aspects of the dragster should be recorded to provide a more complete and detailed history. A possible template for a log book entry is shown on the next page.

the starter series rules

The NXT STEP Dragster Challenge establishes a few rules and guidelines to keep things fun and fair for participants. Each of these rules must be followed to keep it fair and to enter the online Starter Series Records Vault:

* NXT Dragsters must be constructed only from parts found in one LEGO MINDSTORMS NXT #8756 kit. No additional elements (LEGO or otherwise) can be used in this series.
* No altered elements, lubricants, or power systems are allowed.
* The standardized NXT-G program must be used, except that the Light Sensor port connection, the motor directions, and power part connections can be altered in the programming block control panels to match your modified NXT Dragster design.
* The standardized drag strip layout must be used.
* One Light Sensor must be onboard and in a fixed position, facing downward.

* One NXT Brick must be onboard and loaded with six AA batteries of any type. The LEGO NXT rechargeable battery pack is also acceptable.
* Crossing a boundary marker disqualifies the run.

Anything goes within these restraints. And frankly, that gives you a lot of space to explore. More ideas to satisfy your need for speed are lurking in the next section.

going faster starts in the pits

All racers experiment, theorize, upgrade, and test their equipment to improve performance on race day. Experimentation and innovation lead to a faster run down the track. Hitting the drag strip and evaluating the operation of a vehicle is a very important element in success. Now that you've built the NXT Dragster and examined how it performs on the track, it's time to experiment and learn some ways to go faster. There are many improvements that can be made to the NXT Dragster—most important is your ability to identify and eliminate anything that may adversely affect speed. Some necessary compromises will have to be considered to achieve your goal.

Let's consider some things that could be a drag for an NXT Dragster.

Weight

A lighter vehicle means that the power-to-weight ratio is increased by having less weight to carry to the finish line and, hence, a greater acceleration will be achieved, resulting in faster run times. Explore every possible way to cut weight by eliminating unneeded parts, but maintain enough structural integrity to reach the finish line in one piece. Weight distribution is also an important issue. Too much weight over one area, like an axle or one side of the NXT Dragster, can increase resistance, slow it down, or even cause the NXT Dragster to drive off course. Too much weight behind the rear wheels may influence weight transfer to the point of requiring a wheelie bar.

Resistance

This movement-opposing force is created when rotating parts rub against other parts, hindering acceleration and robbing momentum. Be aware of what is on an axle and the spacing between parts, as well as how they perform in action. Too much binding or rubbing on one side of the

LOG BOOK

Elapsed time:
Date:
Time:
NXT Dragster name:
Motors:

* Number used:
* Ports used:
* Motor uses: _____ as drive motors, _____ as special function (explain)

Batteries:

* Six (6) AA Alkaline
* Six (6) AA Rechargeable (type/rating?)
* One (1) LEGO NXT rechargeable battery pack

Wheels:

* Number used:
* Position:
* Used as: _____ drive wheel(s), _____ freewheel(s)
* Type/size:

Gear ratio:
Weight:
Notes:

* Does the NXT Dragster run straight?
* Are any files associated with it (pictures, building instructions, etc.)?
* Are there any special details about the configuration or performance of the dragster?

NXT Dragster can also cause it to drive off course. The elements don't need lubrication and will operate quite efficiently when they are not excessively dirty, loaded by weight, or squeezed or stressed.

Gearing

The gear train is the most important system in the NXT Dragster, and it provides the energy transfer that is needed to reach the finish line. The arrangement of the gears will determine if the NXT Dragster will have more torque or run faster. The NXT Dragster has a gear ratio of 3:1, which means that the 8-tooth gears on the wheel axles rotate 3 times for every 1 rotation of the 24-tooth gears on the motor axles. The motors produce adequate torque; the gear ratios you use will determine how fast your NXT Dragster moves on the drag strip. Torque is needed, but speed is the name of the game. A gear train that provides the optimum balance of torque and speed may start slowly but will accelerate throughout the drag run and, ideally, reach top speed right as the NXT Dragster crosses the finish line. Experiment with different gear configurations and find out which goes faster.

Power

Battery power is very important. Weak batteries will perform poorly. New batteries will perform the best, as will freshly charged rechargeable batteries—although the high-performance edge usually wears off after a few runs, resulting in a speed drop off. Slower elapsed times will indicate that the batteries require replacement or recharging.

further exploration

Racing vehicles are very personalized. One dragster doesn't fit all, and it's the little differences that push the limits of sports and competitors. Whether a dragster is symmetrical or not; has two, three, or four wheels; uses one motor or all three; has a complex gear train or simple direct drive; one thing is for sure—there isn't going to be a standard design in The NXT STEP Dragster Challenge. The only design limits in place are the possibilities that the parts in one NXT kit box can offer. Anything goes from there. The LEGO MINDSTORMS NXT kit is a perfect tool for exploring your potential while learning and having fun in the process. Here are some ideas to stimulate your own creative powers.

* Use other LEGO kit elements. While the most awe-inspiring designs may be one-kit models that actually break records, there's no doubt that elements from other kits offer many opportunities to improve performance. To achieve better run times, you must look at and analyze every physical aspect of your NXT Dragster for possible improvements. Larger rear tires and an enhanced gear train design could help, and weight is a sensitive point you should keep in mind when designing a NXT Dragster.
* Use a motor to shift gears for more speed down the strip—or even to shift into neutral, removing the gear train power transfer for free coasting.
* If you create a one-motor NXT Dragster, use output port A for unbridled horsepower. Port A is not associated with another port, like ports B and C, which are linked, so port A offers greater freedom and optimum performance in single-motor models.
* Use the Ultrasonic Sensor to detect a stop boundary, like a finish flag or an approaching wall. This function is called a *fail-safe* and can be used to stop the NXT Dragster after a run, when crossing a boundary marker, or even after missing the finish line. Using a fail-safe may help you to avoid crashes.
* It is possible to race two NXT Dragsters side by side using two NXT units. They can also communicate via Bluetooth for the start cue. Another idea involves using three NXT units—two as racers and one as a remote race controller that sends the start signal. Both racers could send their ET times back to the controller unit, and the winner could be displayed on the screen.
* Exploring NXT track and race component ideas, such as a speed trap system or a Christmas tree starting light, which can be constructed from the NXT and sensors, will provide many learning opportunities. Complex functions will require some research to build them properly, but the experience you gain will lead to future successes—both on and off the track.

conclusion

You have now gained some experience building, programming, and running an NXT Dragster. You've also learned what you need for entry into The NXT STEP Dragster Challenge; however, this chapter is no place to stop.

Do you feel the need to compete? Check out the growing race scene online at *http://thenxtstep.com/* for interactive coverage, race results, featured stories, custom NXT Dragster designs, and breaking news. If you go to the book section of our forum (*http://thenxtstep.com/smf/index .php?board=8.0*), you will find The NXT STEP Dragster Challenge—Starter Series, where you can compare design notes or discuss the latest in NXT Dragster topics. There you will also find the official Starter Series Records Vault, where all scores will be ranked by elapsed times. If you would like to submit your own results, complete the entry form you find there and send it in. Please be fair and send only those times that were obtained using the exact program code and the specific drag strip layout provided in this chapter.

We want this to be safe and fun for everyone!

5

BobBot: an NXT version of the bobcat

BobBot is a skid-steer loader that is designed to mimic the original Bobcat. With the majority of the weight positioned over the rear wheels, the front wheels barely touch the ground. This weight distribution gives all four wheels good traction while still allowing the robot to steer and turn effectively. We'll use a third motor to grip and lift the load (in this case, a ball).

BobBot's large front door gives you easy access to the NXT Brick for normal operations and battery replacement. The NXT Brick is placed on an angle and held firmly in place by long pins with bushings. The boom is limited in its up-and-down motion by the side pistons, which are not powered as they are in a real skid-steer loader.

By following the building instructions, you'll learn to build three different attachments:

1. A ball gripper, the default model that also is used in the program section

2. A fork for lifting pallets (which mounts on the ball gripper arm)

3. A big grabber, also known as a demolition claw

Figure 5-1: BobBot with ball gripper

BobBot has all four sensors on board. The Light Sensor faces down and can be used for following a line; however, in our program we detect a highly reflective aluminum foil strip to trigger the gripper's lifting mechanism. The Ultrasonic Sensor faces forward to detect and avoid obstacles, the Sound Sensor can be used to start or stop a program section, and the Touch Sensor on the robot's rear will prevent the vehicle from backing into things.

building BobBot

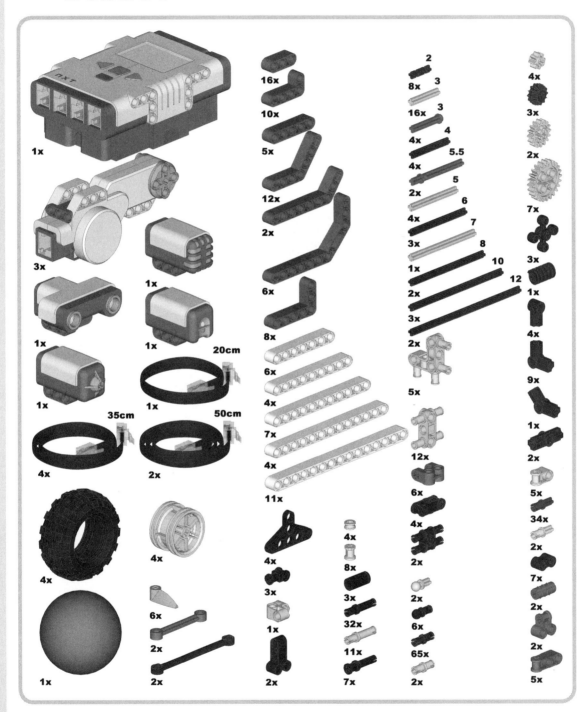

Figure 5-2: Bill of Materials

the right gear base

1

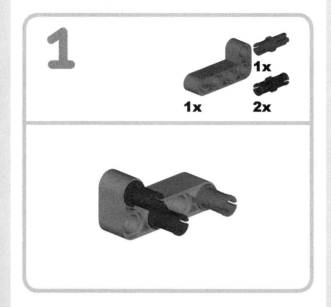

1x   1x   2x

2

1x   1x   1x

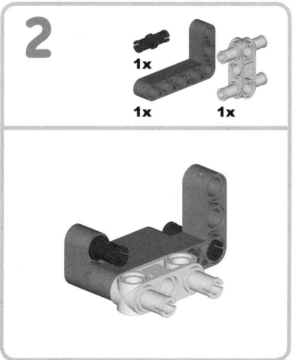

3

2x   1x

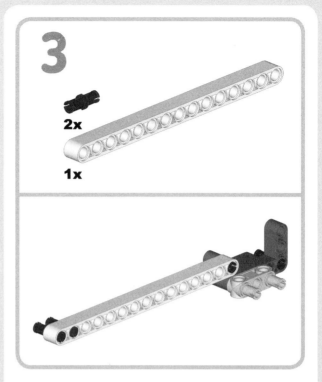

4

2x   2x

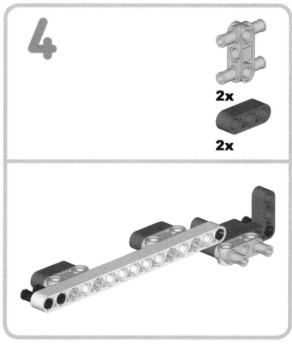

5

1x
2x
2x

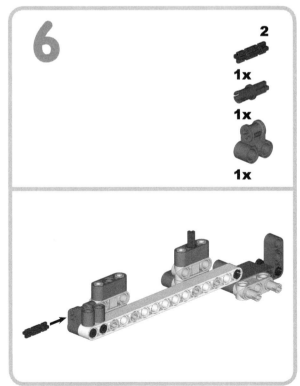

6

2
1x
1x
1x

7

2x
5.5
1x
3
1x
1x

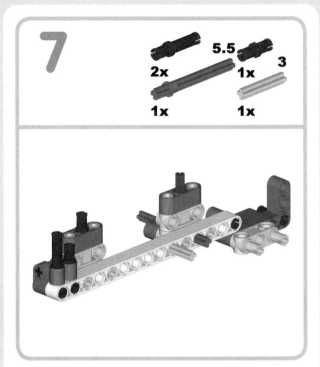

8

1x 2x 1x

9

1x **1x** **2x**

the right base

1

1x

1x

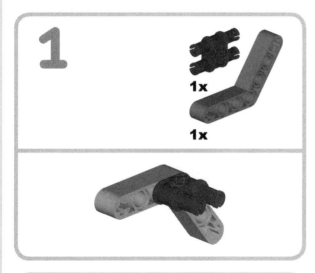

2

1x **1x**

3

1x

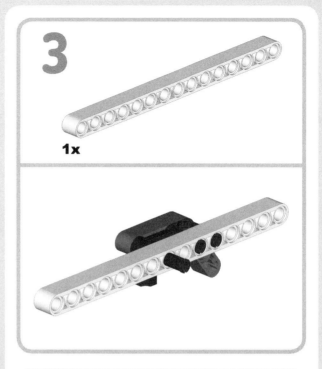

4

1x **1x**

1x **4x**

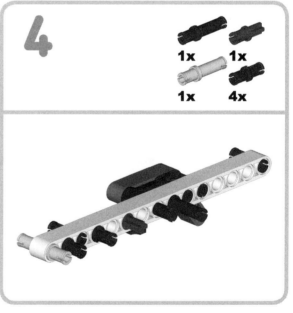

5

1x

1x

1x

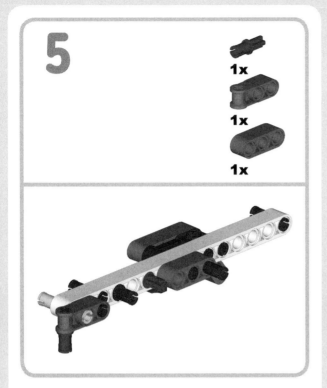

6

2x

1x

2x

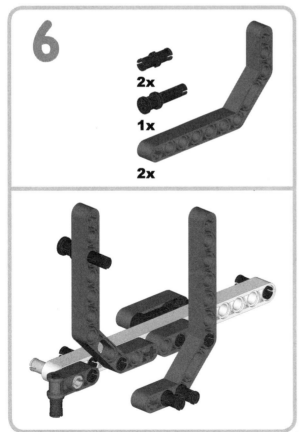

7

1x

1x

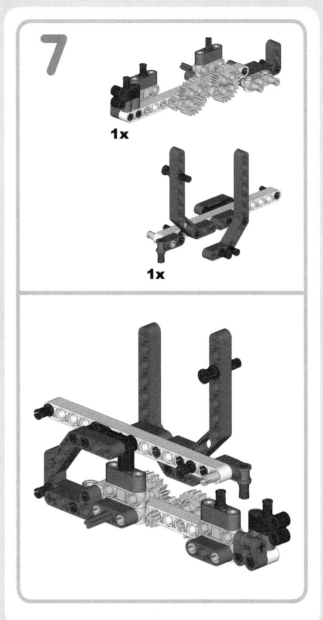

the light sensor

1

50cm

1x   **1x**

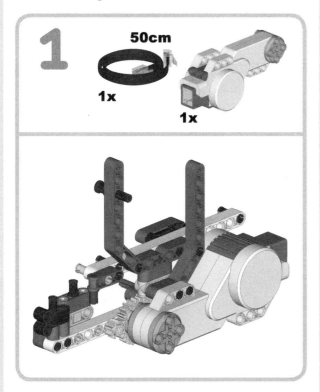

The motor should be connected to the light gray part with two pins; it should also drive the length-3 axle that is inserted in the 24-tooth gear.

2

2x

2x

1x

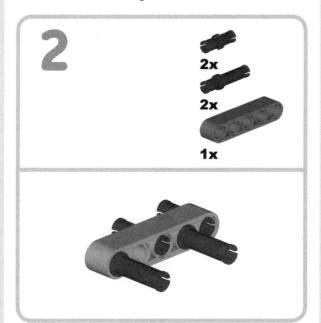

3

4x

1x

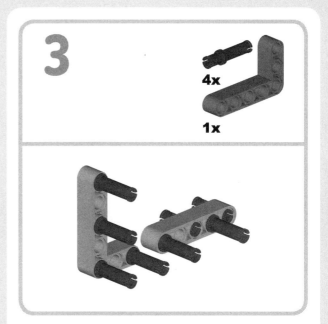

4

20cm

1x

1x   **1x**

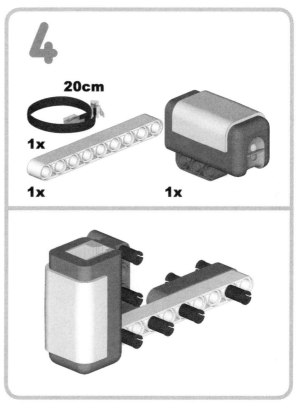

5

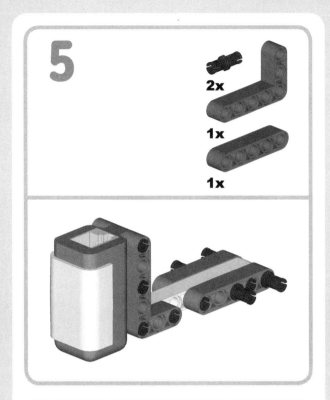

6

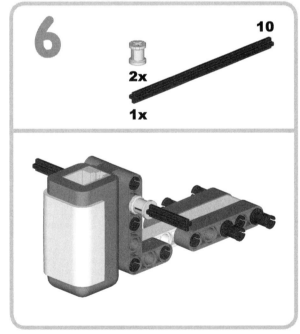

Center the length–10 axle, with a bushing on each side.

7

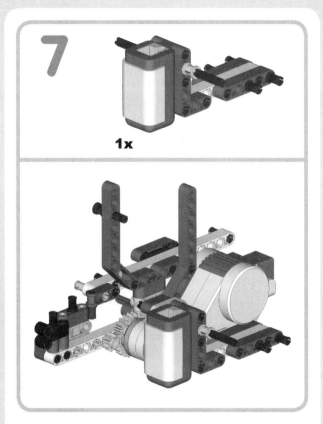

8

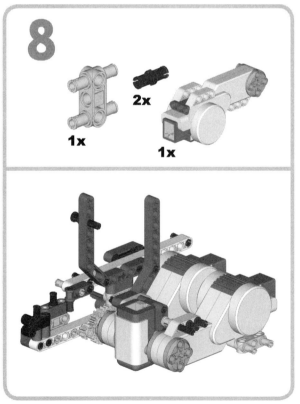

9

50cm

1x

1x

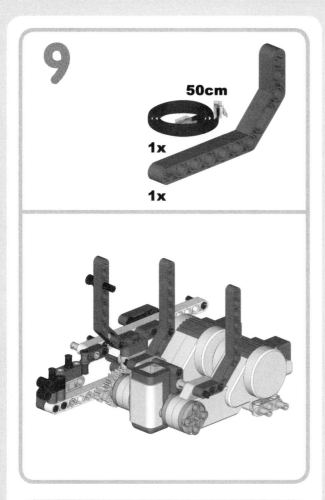

10

4x

1x

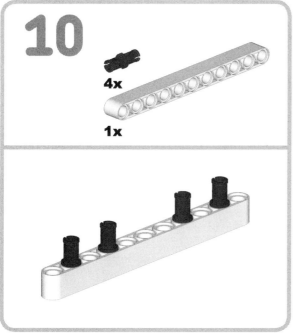

11

1x

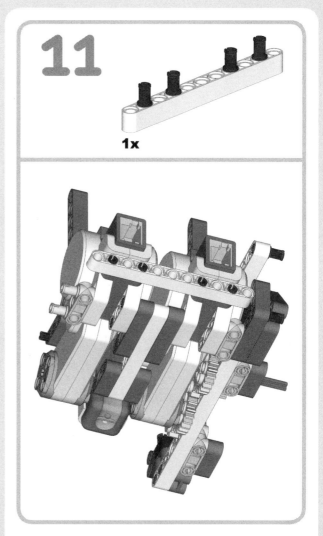

Wrap the two 50-cm/20-inch wires around the beam so they will go inside to the center next to the Light Sensor (see Figure 5-3) on page 149.

the left gear base

1

1x
5.5 1x
1x

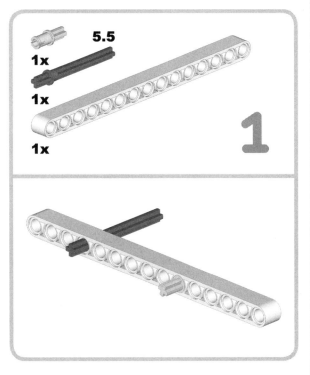

2

2x
3 1x

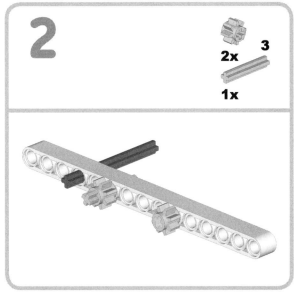

3

1x   1x

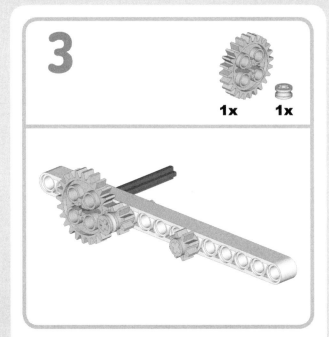

4

3
1x

1x

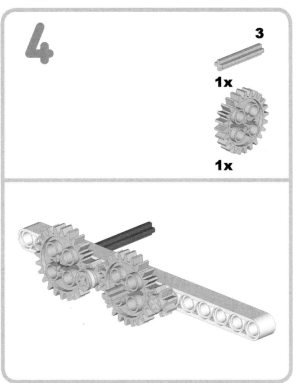

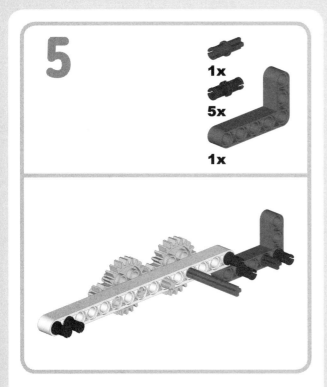

5

1x
5x
1x

The *L*-shaped beam and the length–15 beam are not connected.

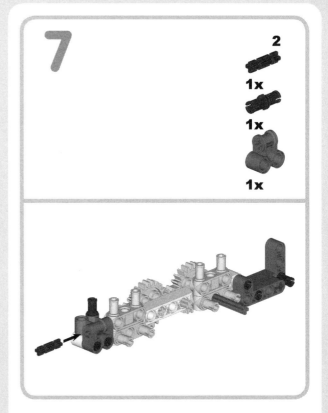

7

2
1x
1x
1x

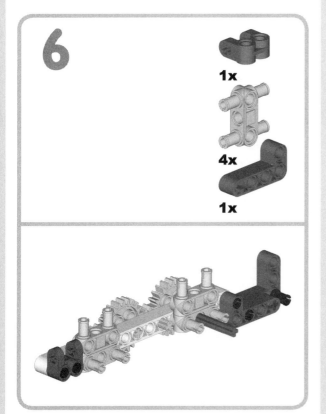

6

1x
4x
1x

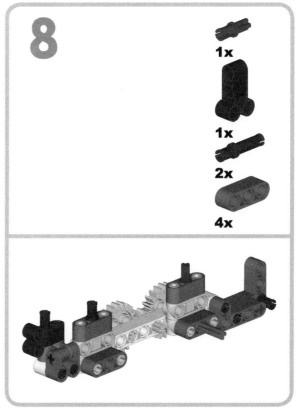

8

1x
1x
2x
4x

the left base

1

1x 1x
1x 1x

2

1x
1x
1x

1x

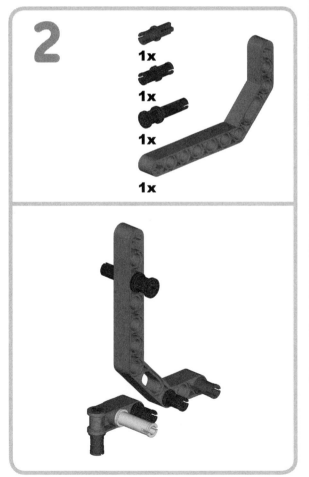

3

2x
1x

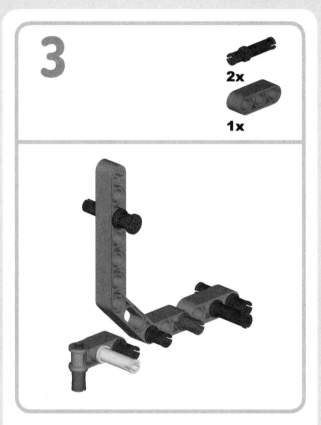

4

1x

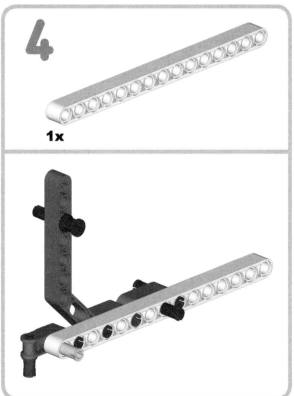

5

2x

1x

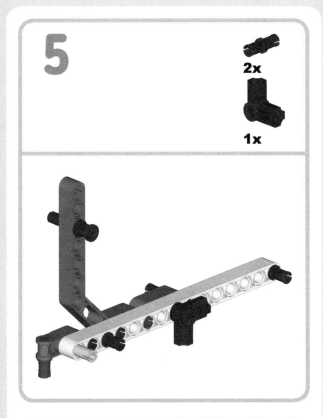

6

1x

1x

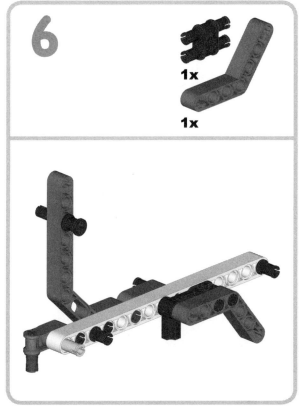

7

1x

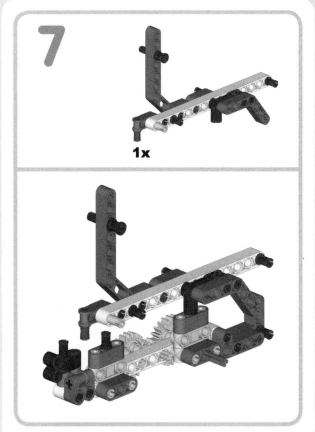

8

1x

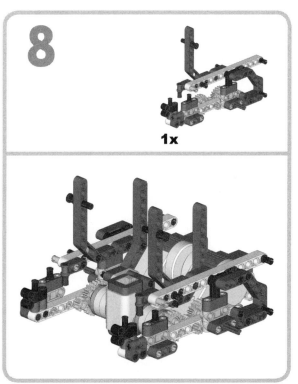

9

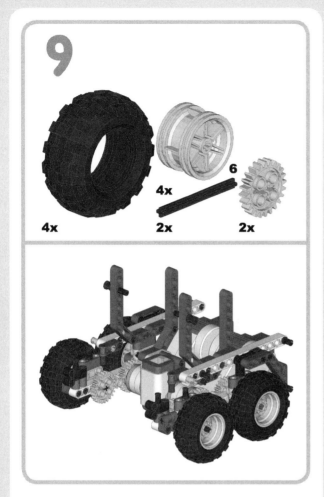

4x   **4x** **6**   **2x**
           2x

Holding the 24-tooth gears, attach the wheels so that the star faces to the inside.

1

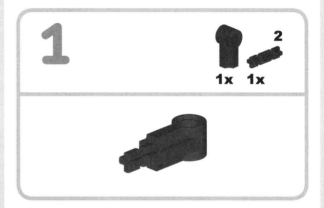

1x **2**
        1x

2

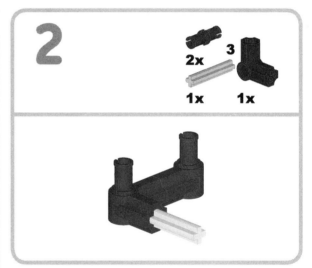

2x **3**
1x **1x**

3

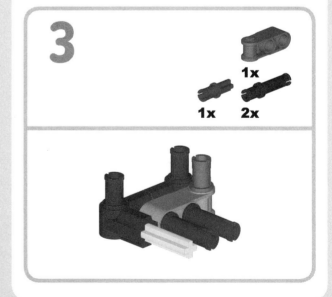

1x
1x **2x**

4

35cm

1x   1x

5

1x

2

1x

1x   1x

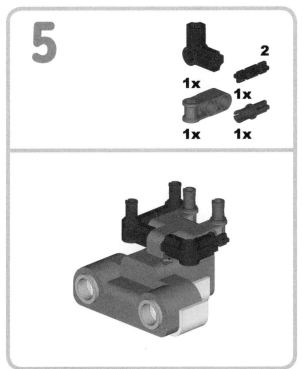

6

1x

2x

7

1x

1x

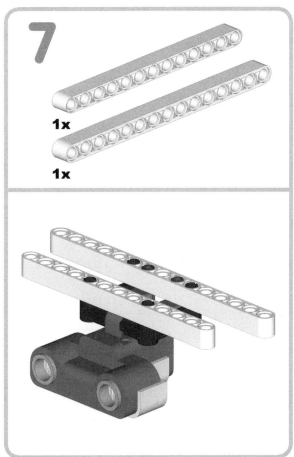

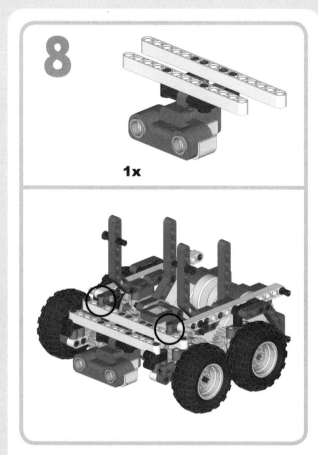

8

1x

Don't forget to place the top blue axle pins in the white length-15 beam on the Ultrasonic Sensor unit.

test BobBot's base

Before you test the robot, connect the wires from the sensors and motors to the NXT Brick. Connect the short (20 cm/8 inch) wire from the Light Sensor to input port 3, and connect the medium (35 cm/14 inch) wire from the Ultrasonic Sensor to input port 4. Use long (50 cm/20 inch) wires to connect the right motor (the first one you placed) to output port B and to connect the left motor (the second one you placed) to output port C.

Figure 5-3 shows how to place the wires in between the base and the NXT Brick. Placing them in a loop makes it easier to attach the Brick to the driving base. Use the two pins with bushings that you've already inserted to secure the Brick.

At this stage, it is a good idea to test the base. The NXT Brick comes with a programming function that allows us to program from the Brick itself. (To learn more, see pages 14 and 15 of the manual that came with your NXT kit.) Follow these steps to enter this programming mode:

* Press the orange button to turn on the NXT Brick, then press the right gray arrow button until the screen displays *NXT Program.*
* Press the orange button to select this mode. Make sure all wires are connected.
* Press the orange button again to start programming.

Now start programming by using the left and right arrow keys to select a desired action or command. The orange key takes you to the next step, and the dark gray key will undo the previous step. Table 5-1 lists the steps needed to create a program that makes the base avoid objects or walls.

Test the program after you have entered all the steps. To do so, place BobBot on the floor about 30 cm (12 inches) from a wall, then run the program by pressing the orange button. The BobBot base should drive toward a wall and then turn to avoid the wall. Note that while the program statement is *Turn Right*, the robot should turn left based on its gearing.

Now we'll continue assembling the cabin and add two more sensors and the third motor. For ease of construction, temporarily remove the NXT Brick from the base (which may require you to disconnect the wires from the Brick).

Figure 5-3: Wire placement for BobBot base with NXT Brick added

Table 5-1: Test Program Steps

Step	Action	
1	Forward	(Unlimited)
2	Object	
3	Turn	Right
4	Wait	2
5	Loop	

the left side cabin

1

2
1x
1x
1x

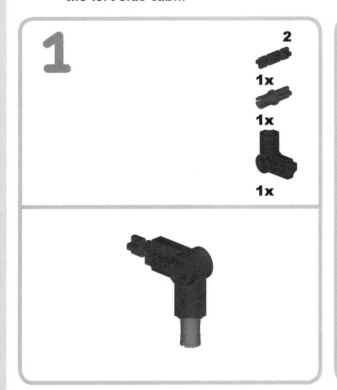

2

1x
1x
1x

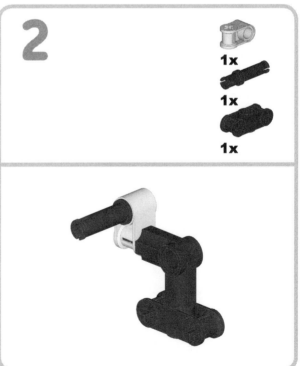

3

1x
1x

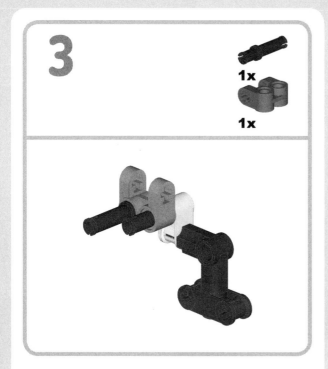

4

1x
1x

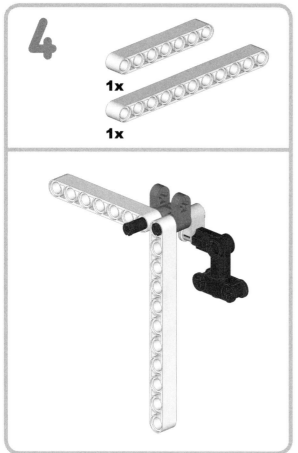

5

3x
2x

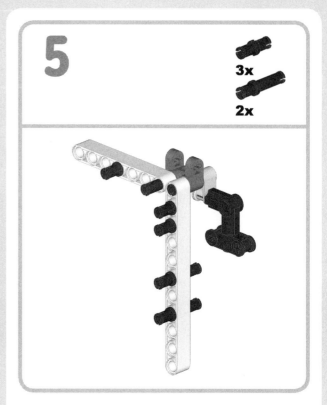

6

2x
1x

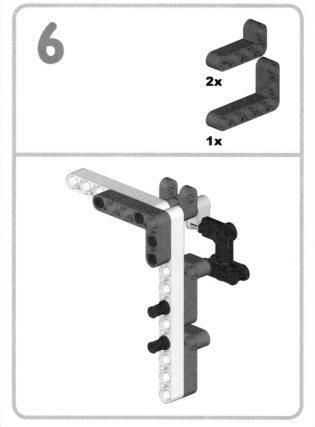

7

2x **1x** **8**

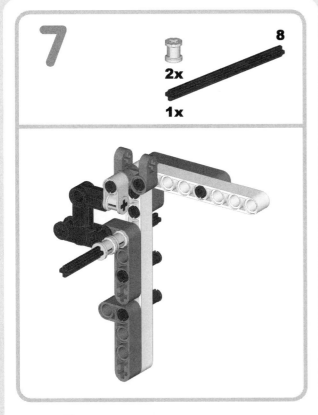

the worm gear

1

1x **1x** **2x** **6**

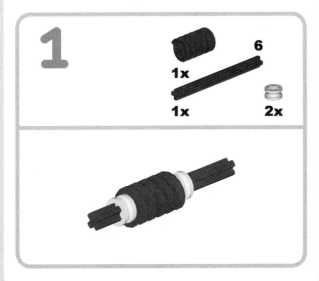

2

1x **2x**

3

1x **3x** **3** **4**

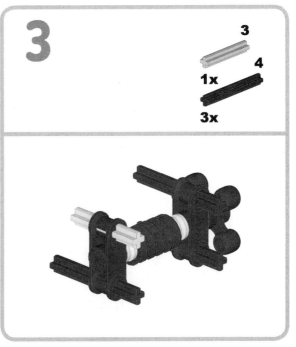

4

4x

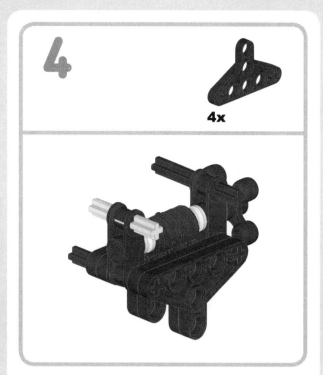

Place the axles exactly as shown. They must not stick out any farther.

5

1x

1x

6

1x   1x

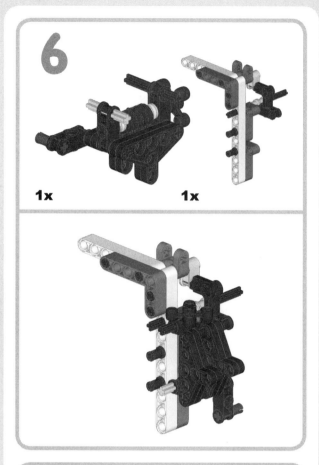

7

1x

1x

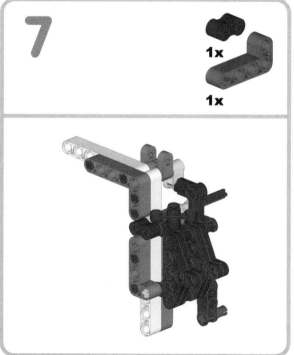

the sound sensor

1

1x

1x

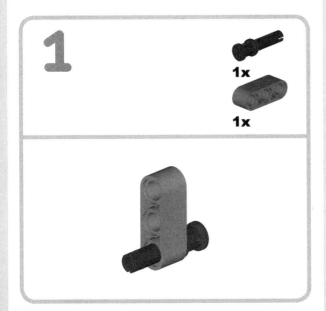

2

3

1x

35cm

1x

1x

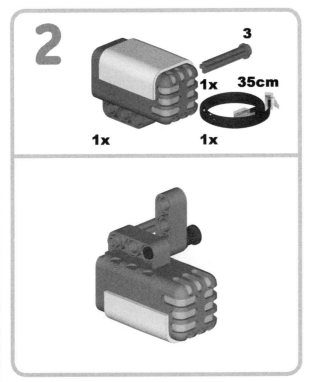

3

1x

1x

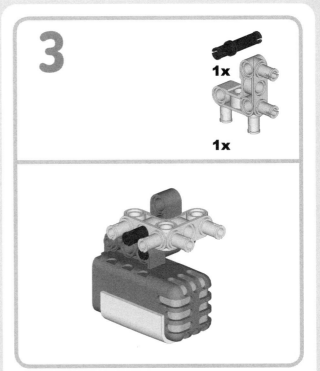

4

1x

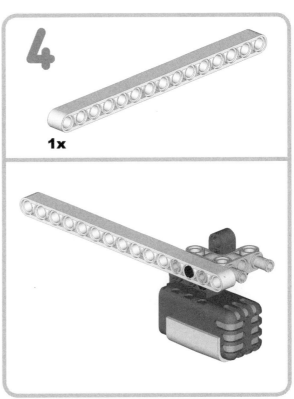

5

2x
1x   **1x**

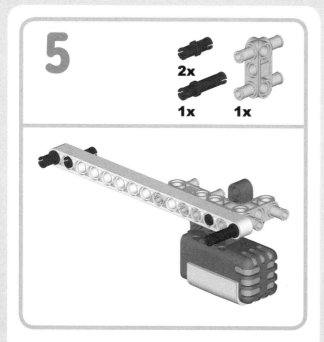

6

1x

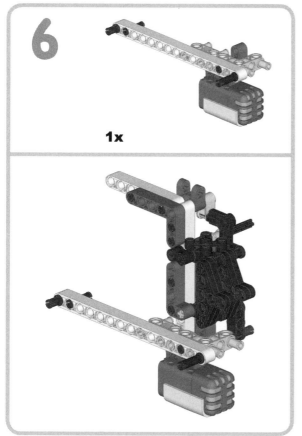

The length–3 axle with the stud goes in the last hole of the vertical cabin beam.

7

1x

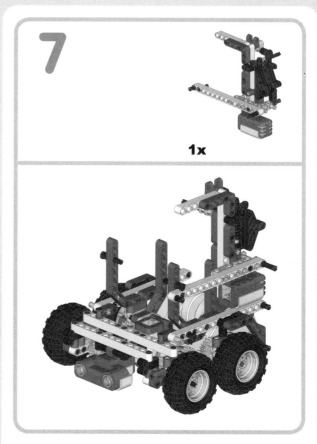

8

1x
1x

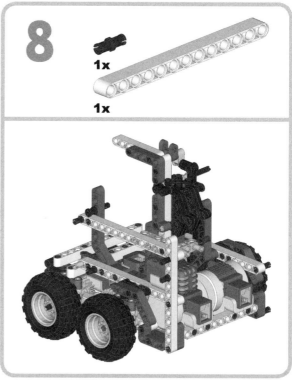

the arm motor

1

35cm

1x

1x

First connect one end of a 35-cm/14-inch wire to the motor. It is best if the natural curve of the wire is toward the front side (inside).

2

2x
5
1x

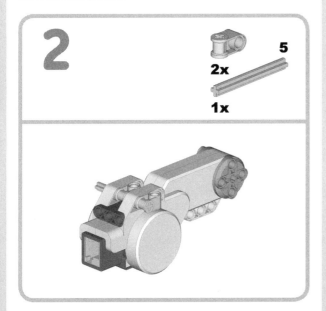

3

1x
1x

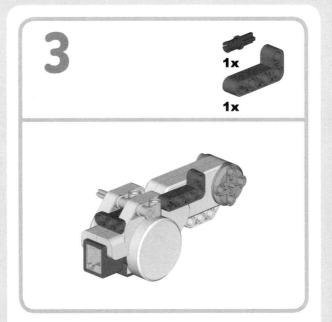

4

2x   1x

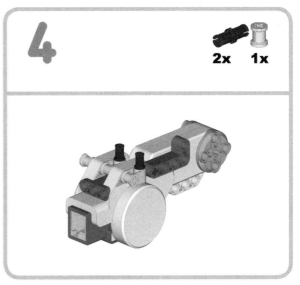

the rear fender

1

1x 1x 1x **3**

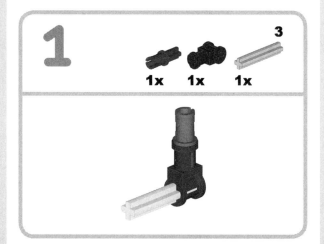

2

1x 1x 1x

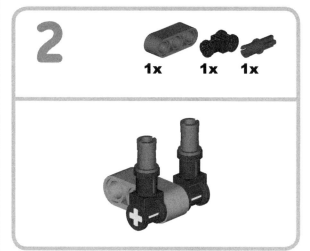

3

1x **3** 1x

1x 1x

4

12

1x 2x

1x 2x

Center the length-10 axle.

5

1x

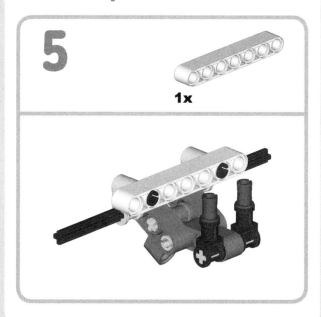

6

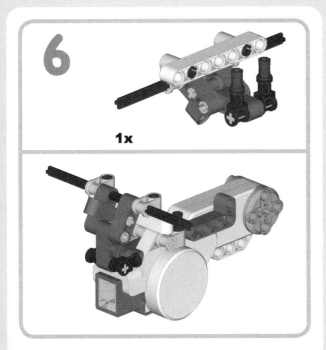

1x

7

1x

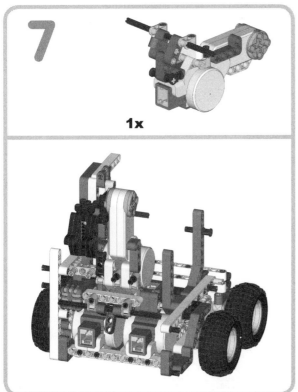

Add the motor subassembly by sliding the axle through the middle of the three holes on the top of the motor, and insert the blue axle pin in the center hole of the bottom motor support.

8

1x

7

1x   2x

9

1x

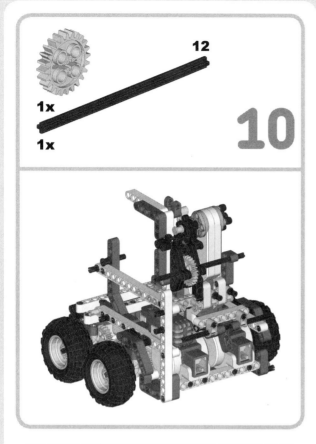

10

12
1x
1x

11

5
1x
1x

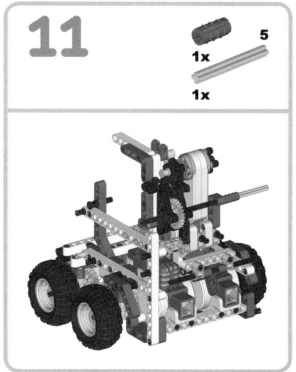

12

1x
1x

The gray bent beam slides on the length-10 axle; make sure to hold the white beam that is also fixed to this axle so that it is facing toward the NXT motor.

Place the white beam that faces the motor inside and on top before you place the gray bent beam.

13

3x

14

1x
1x

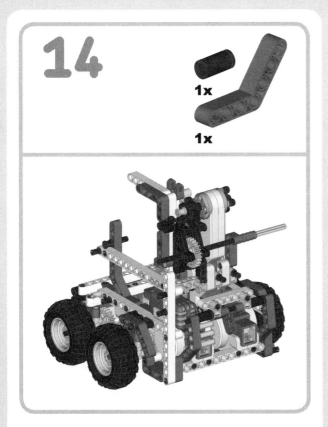

15

3x

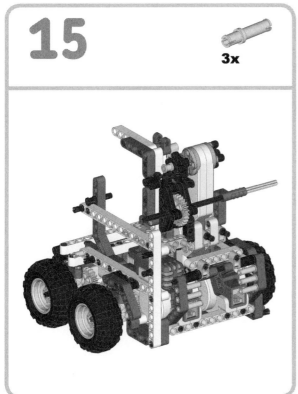

16

3
1x

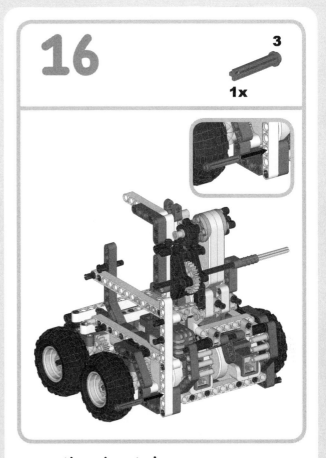

the exhaust pipe

1

3
2x
1x

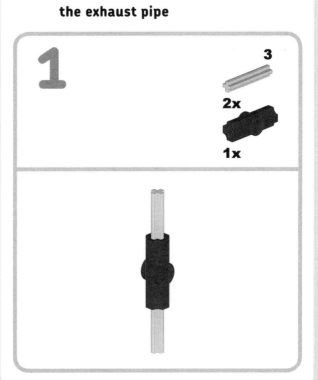

2

1x 1x

5

1x

the touch sensor

3

2

1x 1x 1x

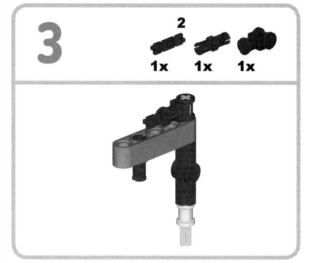

1

1x 1x 1x

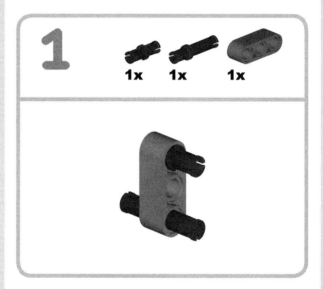

4

1x 1x

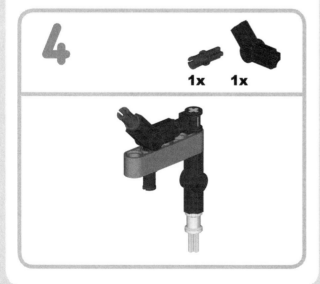

2

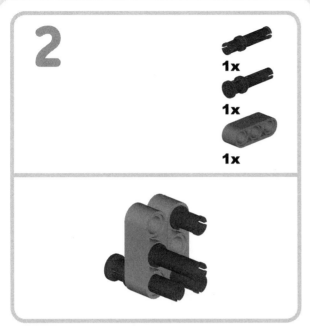

1x

1x

1x

3

35cm

1x

1x

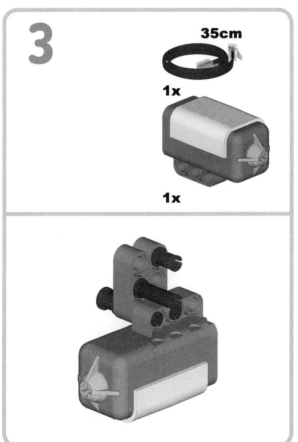

4

2x

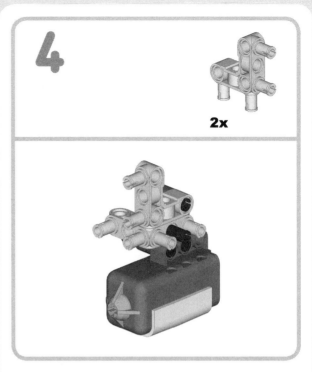

These two pins are referred to as *Hassenpins*. Connect them to each other before adding them to the robot.

5

1x

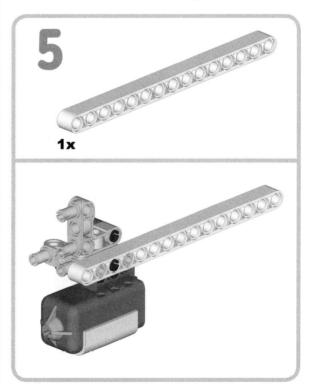

6

1x

1x

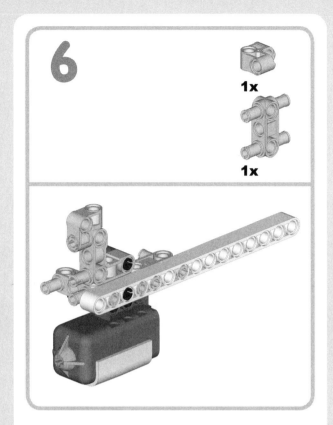

7

2x

1x

8

1x

1x

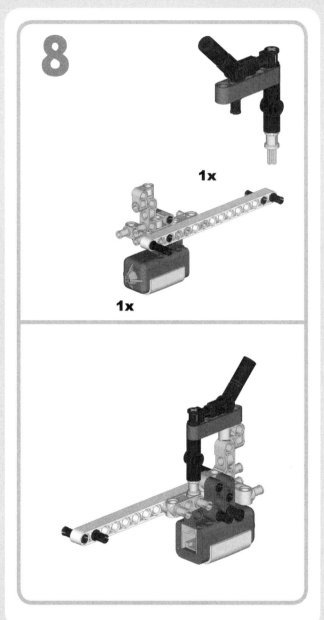

the right cabin

1

2x
2x
1x

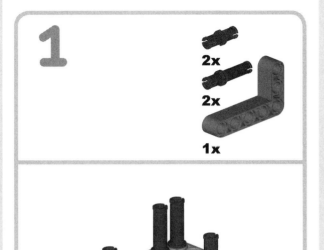

2

1x
1x

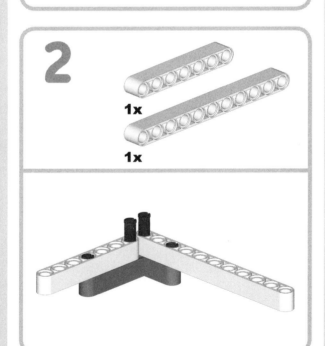

3

1x
1x

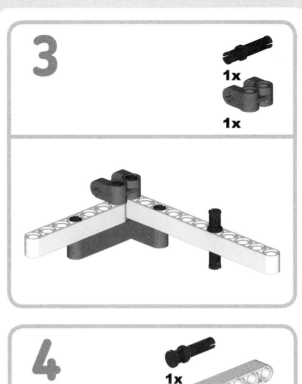

4

1x
1x

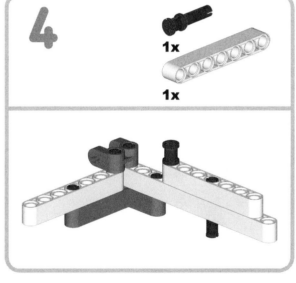

5

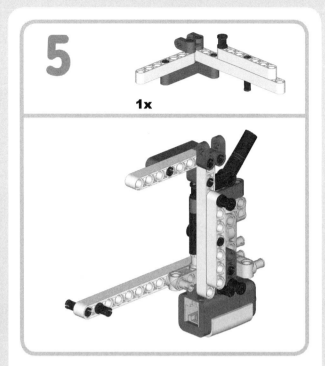

1x

6

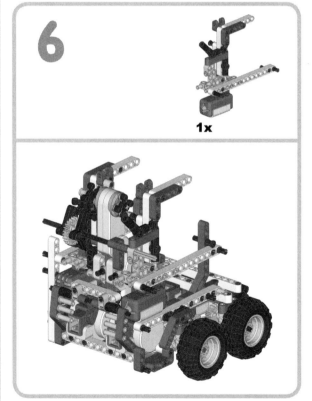

1x

The black pin with a bushing goes on the length-10 axle that extends from the motor. Keep the Touch Sensor wire on the inside of the cabin.

7

3

1x

1x

1x

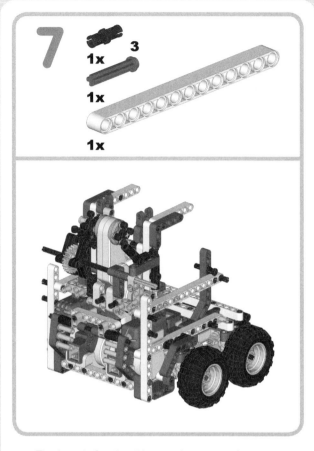

The length-3 axle with a stud connects the rear fender to the main body.

the rear lights

1

2x

1x

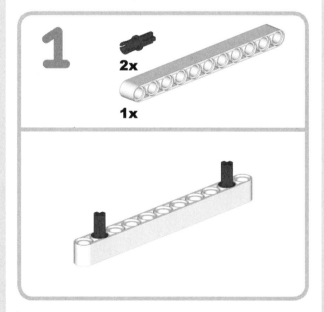

2

2x

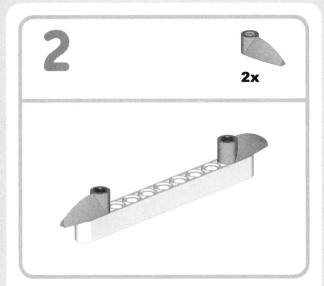

3

1x

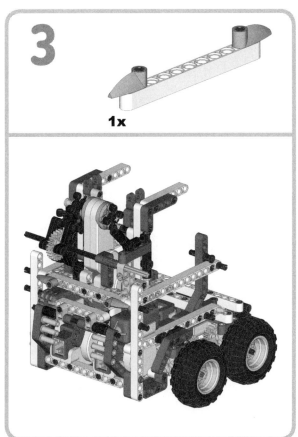

4

1x

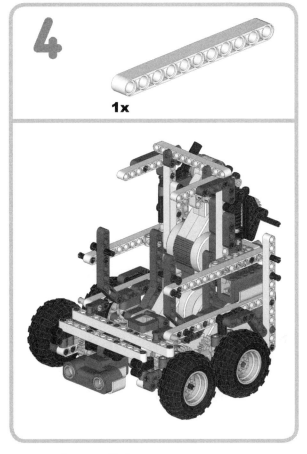

the top lights

1

2x   3   2x

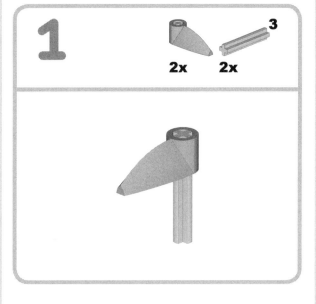

Make two of these.

2

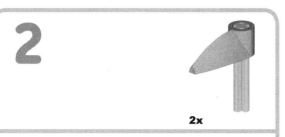

2x

connecting the cables

Connect all the wires to the Brick as shown in Table 5-2.

Table 5-2: Wire Placements for BobBot

Wire length (cm/inch)	From motor/ sensor	To NXT Brick port
Short (20 cm/8 inch)	Light Sensor	Input port 3
Medium (35 cm/14 inch)	Ultrasonic Sensor	Input port 4
Medium (35 cm/14 inch)	Touch Sensor	Input port 1
Medium (35 cm/14 inch)	Sound Sensor	Input port 2
Medium (35 cm/14 inch)	Arm motor	Output port A
Long (50 cm/20 inch)	Right motor (first placed)	Output port B
Long (50 cm/20 inch)	Left motor (second placed)	Output port C

Insert the NXT Brick from the front, and make sure that all the wires are nicely packed underneath. If you are using a USB cable to download the programs to your robot, take the Brick out of the robot to connect the USB cable.

1

1x

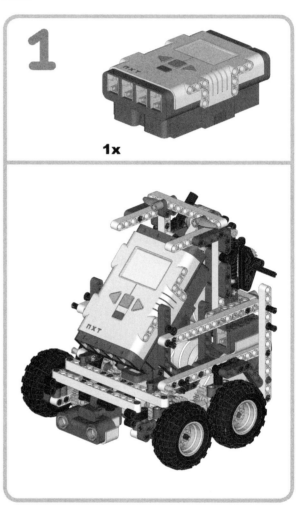

the cabin door

1

3x

2x

1x

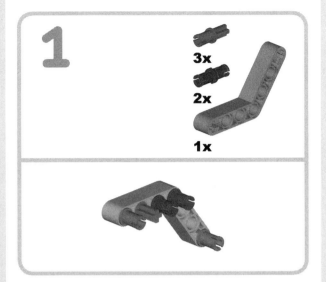

2

1x

1x

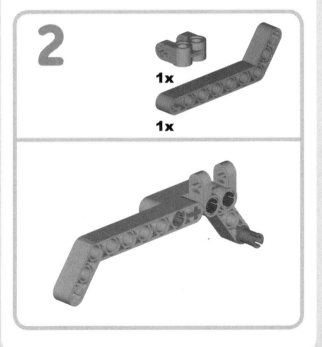

3

1x

1x

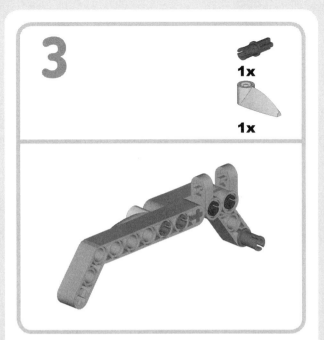

4

10

2x

1x

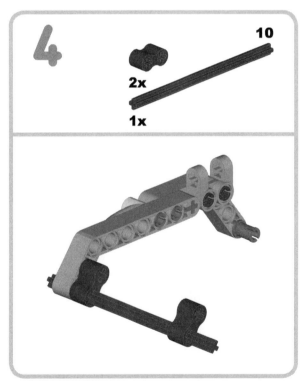

5

2
1x

1x

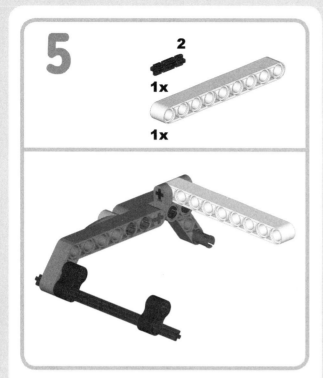

6

2
1x

2x

1x

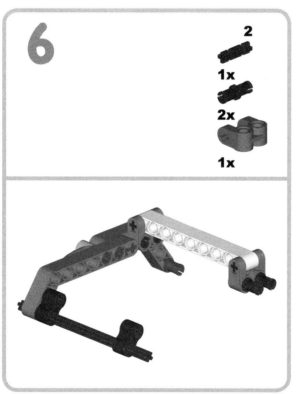

7

3x

1x

8

1x

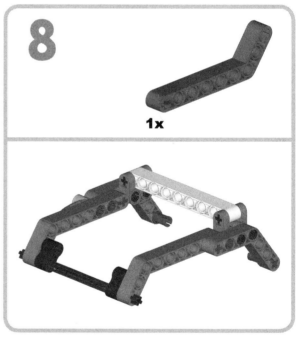

9

1x

1x

10

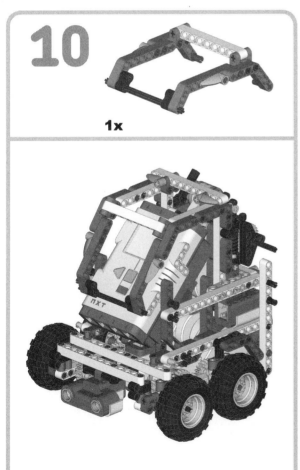

1x

the arms

1

1x

1x

1x

2x

1x

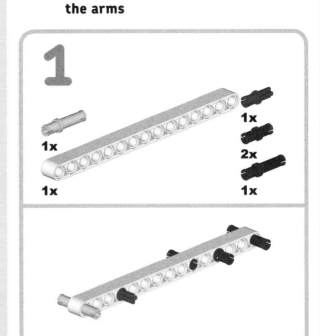

2

1x

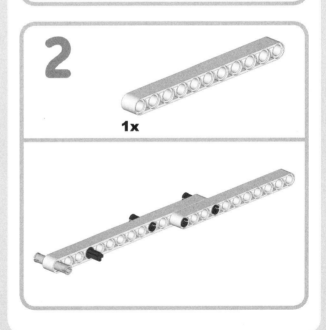

3

1x

1x

1x

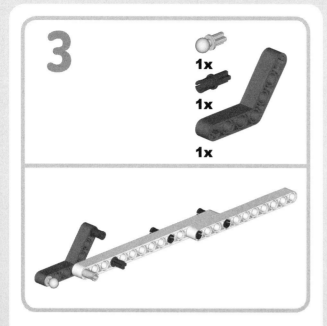

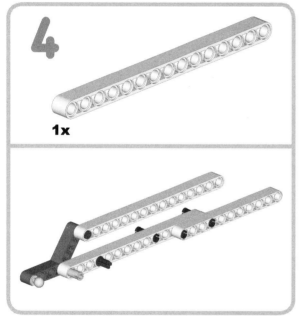

4

1x

5

2x

1x

6

1x

7

1x
1x

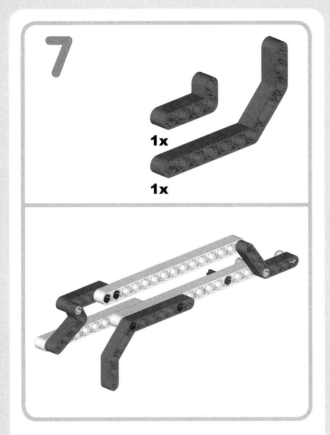

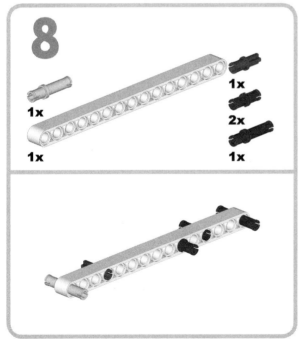

8

1x
1x
1x
2x
1x

9

1x
1x
1x

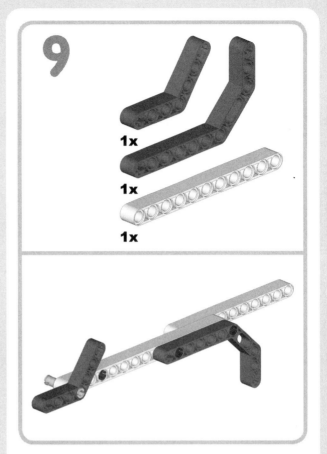

10

1x
1x

11

1x

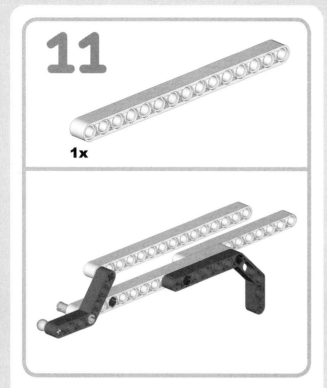

13

1x
1x

the ball gripper

12

2x
1x

1

2x
1x

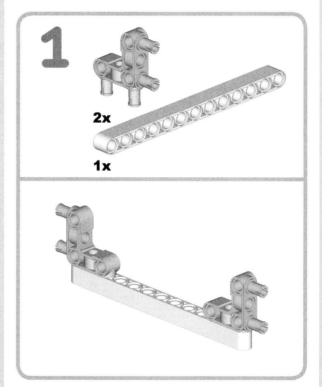

2

2x **3**

1x **4**

1x

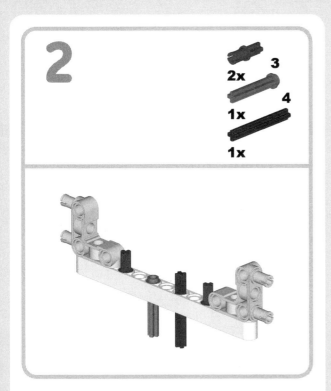

3

1x

2x

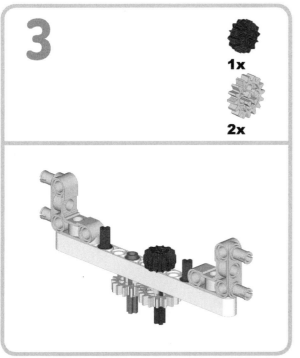

4

2x

2x

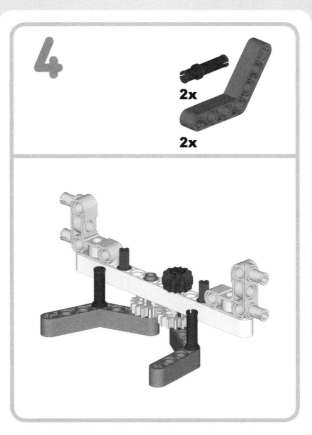

Put both grippers in the outermost position.

5

2x

8

1x

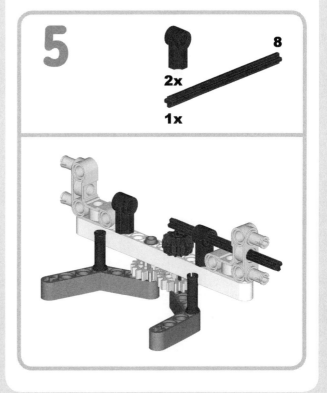

6

1x
1x

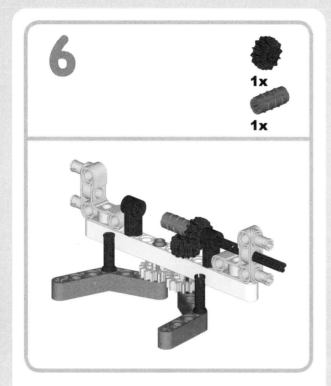

7

1x
1x
10

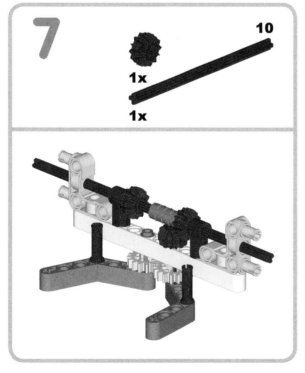

8

1x
1x

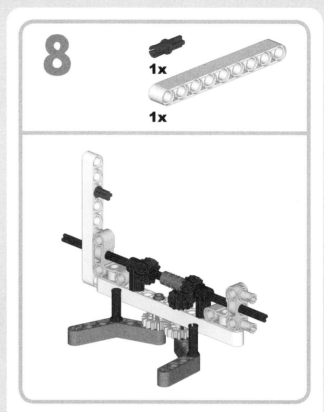

9

1x
1x

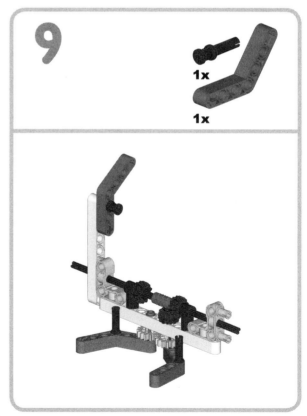

10

1x
2x
5

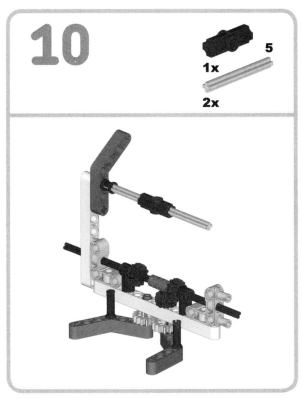

11

1x
1x
1x

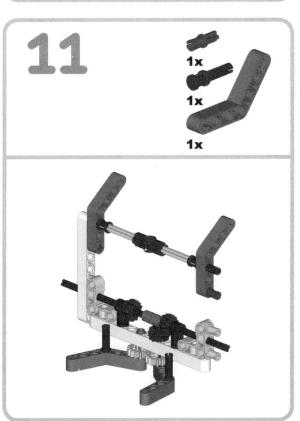

12

2x
1x

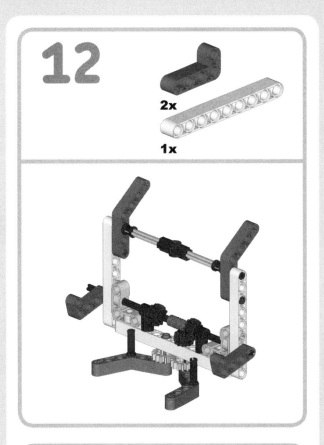

13

2x

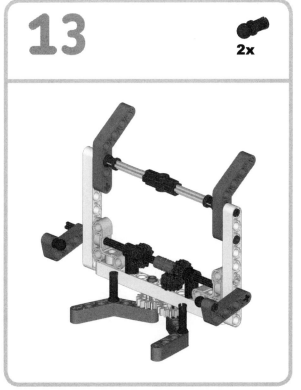

14

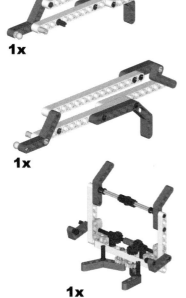

1x

1x

1x

15

2x

16

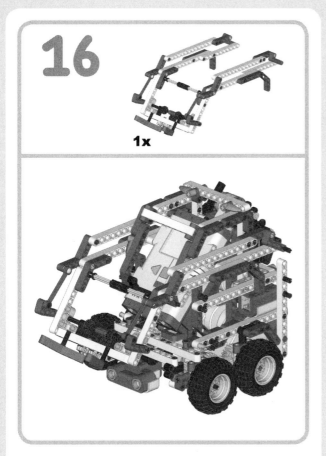

1x

Note that both arms are placed at the same height before placing the four beams on the axle. You may need to turn the knob wheel to adjust the lifting axle.

17

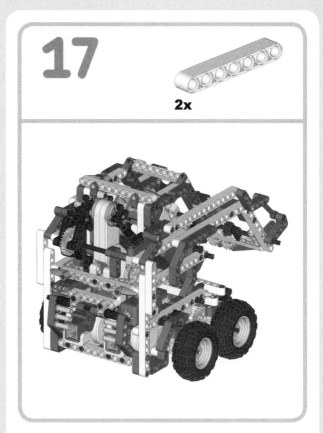

2x

18

2x

1x   1x

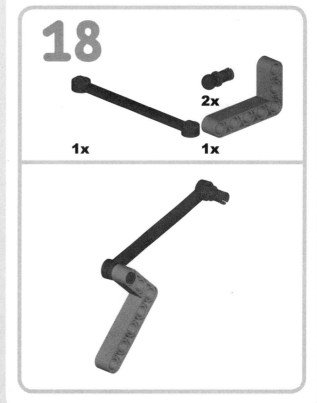

19

1x

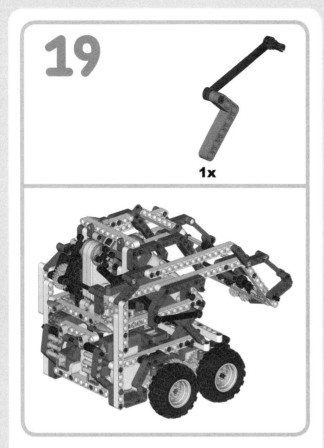

20

2x

1x   1x

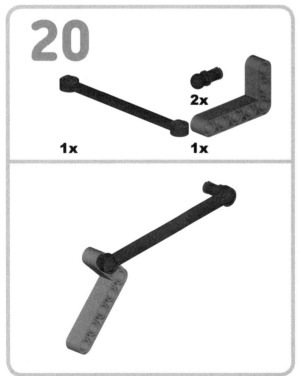

21

1x

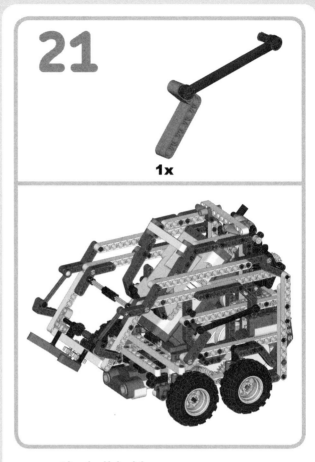

the ball holder

The ball gripper can only grab the ball if it is placed on the ball holder.

1

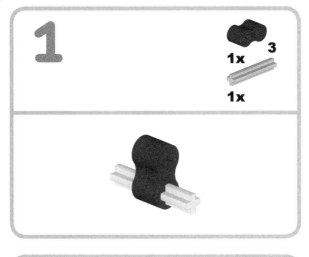

1x 3
1x

2

2x

3

2x 3
2x

4

1x 2x 1x 3

Add the ball, and your BobBot is ready to go!

5

1x

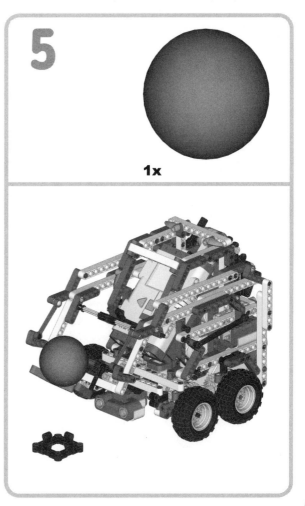

optional accessory building instructions

Like a real skid-steer loader, BobBot is versatile, and various other attachments can be built with just one kit. Here are a few ideas to get you started.

fork

It is easy to add a fork to BobBot. To do so, follow these instructions.

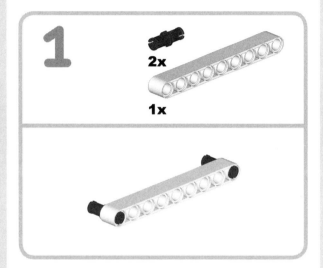

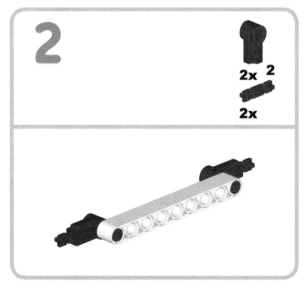

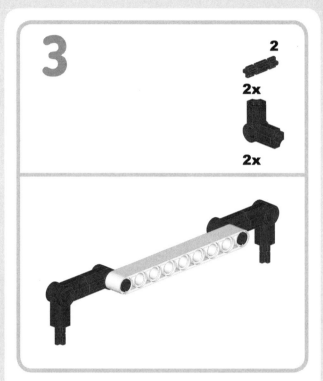

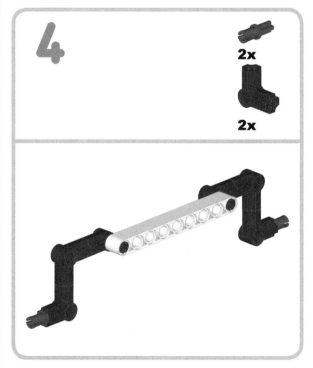

5

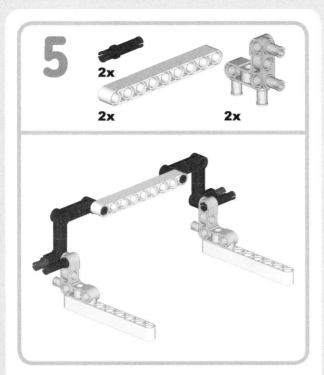

When you've built the fork, connect it to the ball-grabbing arm as shown in Figure 5-4.

Figure 5-4: BobBot with ball gripper and fork

demolition claw

Another optional attachment (which can also be constructed with just the base kit) is a big demolition claw. Before we build it, we need to disassemble the ball gripper arm (see Figure 5-5) because we need to reuse its parts. We can keep the arms and attach the claw to them.

Figure 5-5: Remove these parts from BobBot's ball gripper.

1

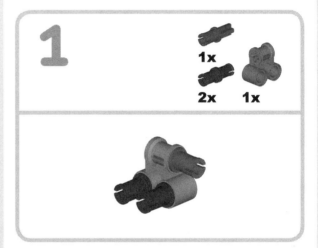

2

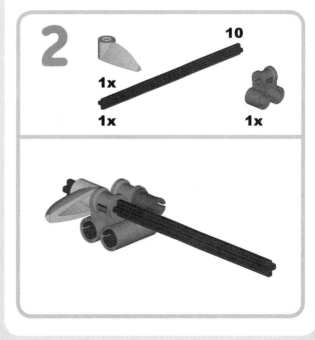

3

1x

1x   1x

5

1x   3

1x

4

1x

1x

1x

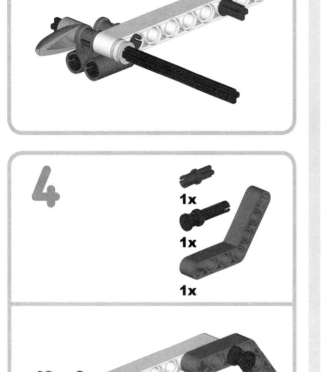

6

1x

1x

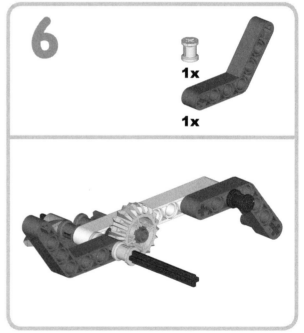

7

1x
1x

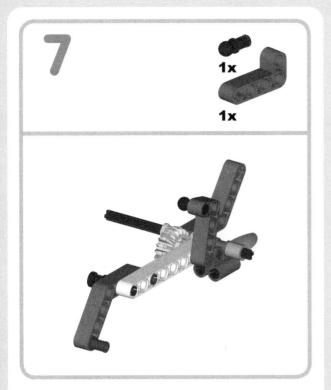

8

1x
2x

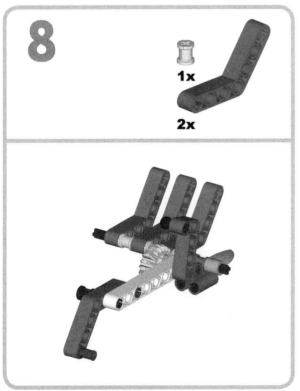

9

10
1x
1x

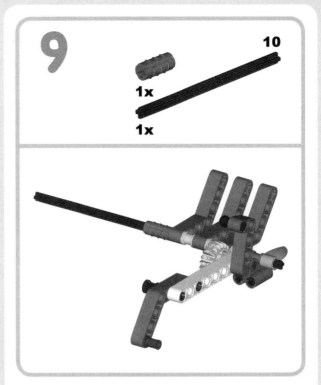

10

1x
2x
3x

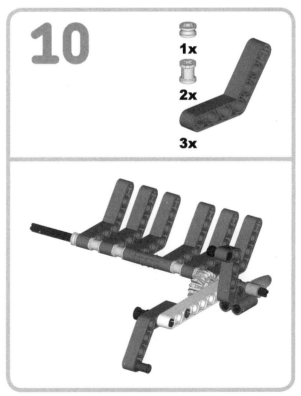

11

1x
1x
5
2x

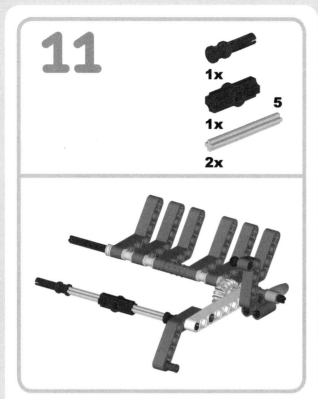

12

1x
1x

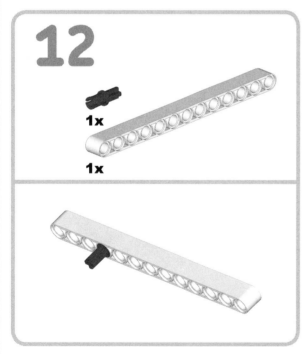

13

1x
5
1x
1x

14

1x
1x

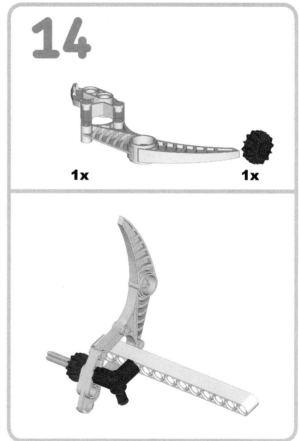

15

1x 1x 7

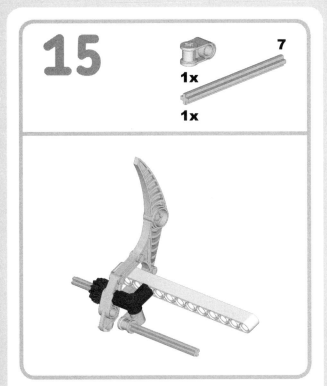

16

2x

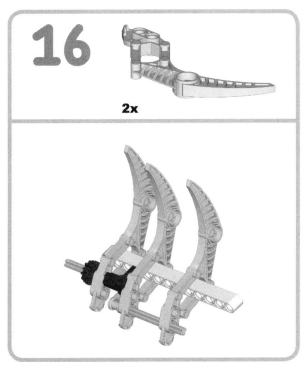

17

1x

18

1x 5 1x
1x

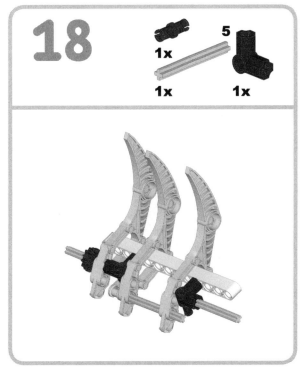

19

1x

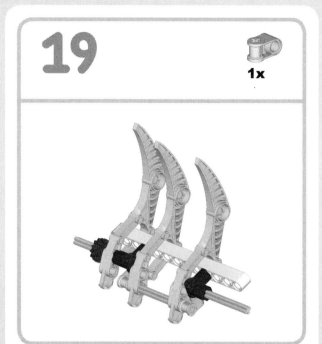

20

1x   1x

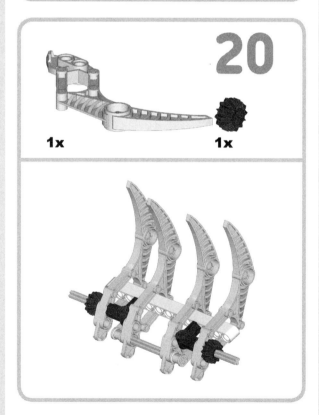

Insert the claw in the most open position; the white beam should rest on top of the gray bent beam. At the same time, make sure that the length-4 beam is pushed down as far as possible.

21

1x

1x

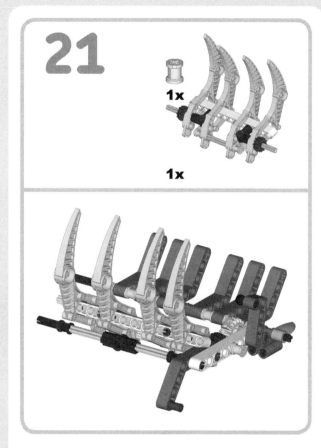

22

2x

1x

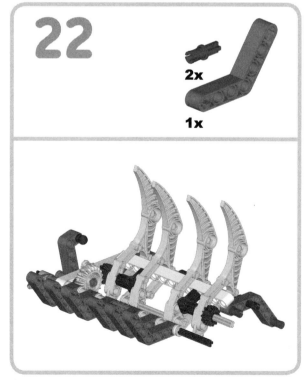

23

1x · 1x · 1x

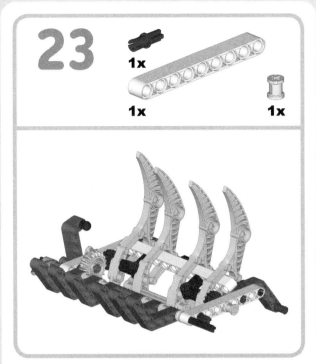

25

1x · 1x

24

1x · 1x · 4

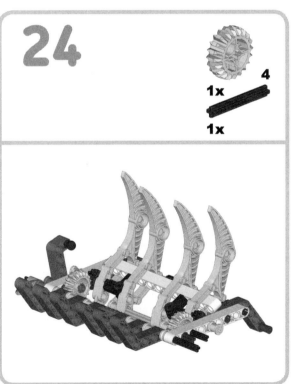

26

2x · 2x · 1x

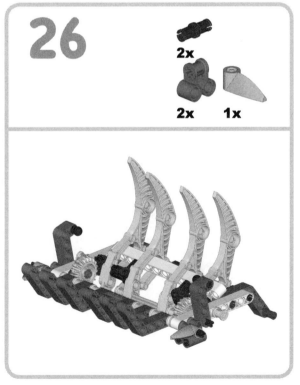

Place the gear and the length-4 beam at the same angle as the other length-4 beam.

Place the claw on the arms, as shown in Figures 5-6 and 5-7. Note the open hole on the gray beam.

Figure 5-6: The claw placed on BobBot's arms

Figure 5-7: BobBot with demolition claw

programming BobBot

This program is written for the basic BobBot with the ball gripper. Using other attachments will require you to modify the program.

Spread the test pad from your NXT kit on a flat surface, place the ball and its holder in the center of the red ball location on the test pad, and then position BobBot behind the starting line. Place a bucket or small low bin (I built one with standard 2-by-4 bricks) on the right side of the test pad.

To create a trigger for the Light Sensor, cut a 2.5 cm (1 inch) wide strip of high reflective aluminum foil, and place it about 2.5 cm (1 inch) in front of number 180 on the test pad; the strip should be about 14 cm (6.5 inches) in front of the ball.

NOTE The robot cannot use its Ultrasonic Sensor to detect the ball because the distance to a round ball is hard to measure and because the gripper is positioned in front of the Ultrasonic Sensor.

Before BobBot can work independently, the gripper must be in the lowest open position. Because the gears make it hard to turn the arm by hand, use the left and right arrow keys on the NXT Brick to raise and lower the arm (motor A). The rest of the programmed actions begin when the orange (Enter) key is pressed for the second time.

BobBot starts moving slowly forward until it detects the strip. It is not yet at the ball, so we need to drive forward for a predefined time and then stop to pick up the ball. After the gripper grabs and lifts the ball, the vehicle makes a half turn to the right and scans for the bucket, where it should dump the ball. The Ultrasonic Sensor detects the distance to the bucket and drives forward until the gripper is over the bucket. Then the arm is lowered, and the gripper will rest on the edge of the bucket; this will open the gripper and release the ball. Finally, the robot raises the gripper and backs up away from the bucket. Figure 5-8 shows the complete program.

The following figures provide an overview for raising and lowering the gripper with the arrow keys and should be used at the start of the program.

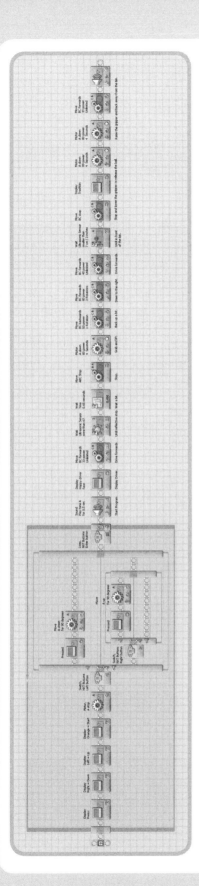

Figure 5-8: Complete ball-gripper program

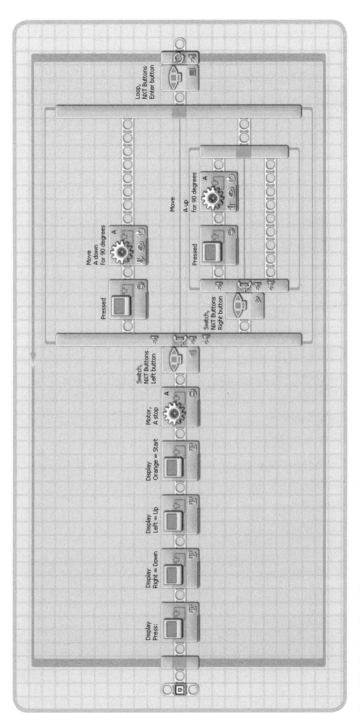

Figure 5-9: Use the left and right arrows to lower and raise the arm.

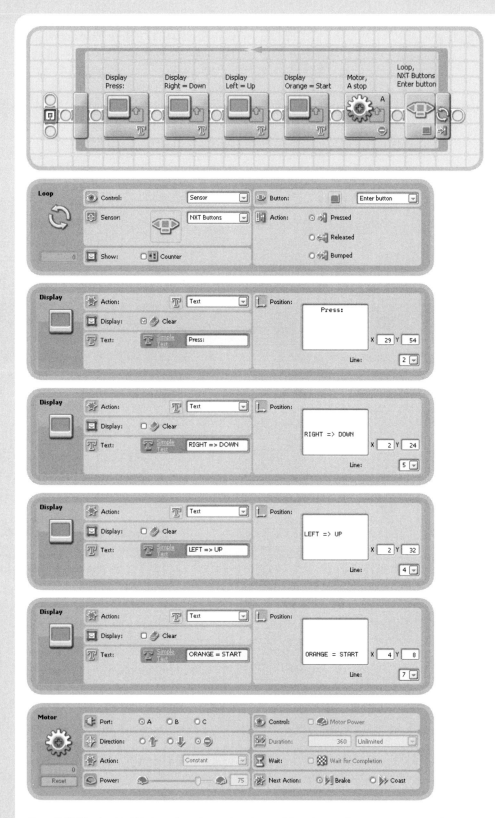

Figure 5-10: The six added blocks perform these actions: loop until Enter button, display text, stop motor A

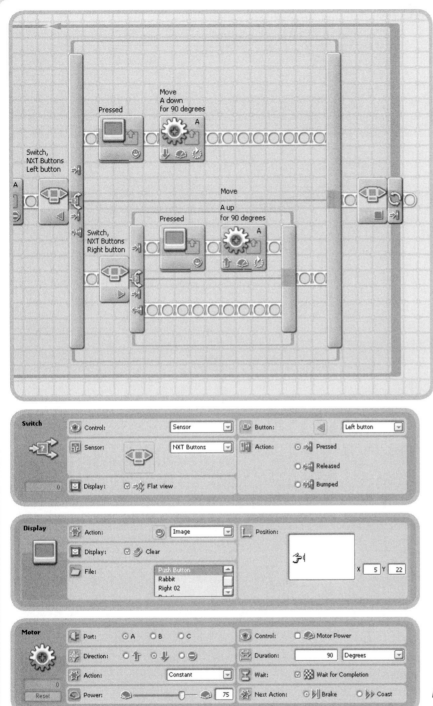

Switch

Control: Sensor
Sensor: NXT Buttons
Display: ☑ Flat view   0

Button: Left button
Action: ◉ Pressed
○ Released
○ Bumped

Display

Action: Image
Display: ☑ Clear
File: Push Button / Rabbit / Right 02

Position:
X 5 Y 22

Motor

Port: ◉ A ○ B ○ C
Direction: ○ ↑ ◉ ↓ ○ ⊖
Action: Constant
Power: 75   0   Reset

Control: ☐ Motor Power
Duration: 90 Degrees
Wait: ☑ Wait for Completion
Next Action: ◉ Brake ○ Coast

Figure 5-11: The left and right arrow keys trigger movement.

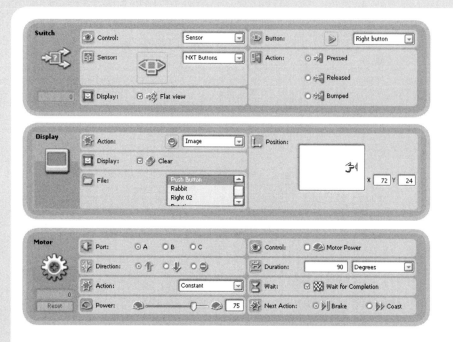

Figure 5-11 (continued): The left and right arrow keys trigger movement.

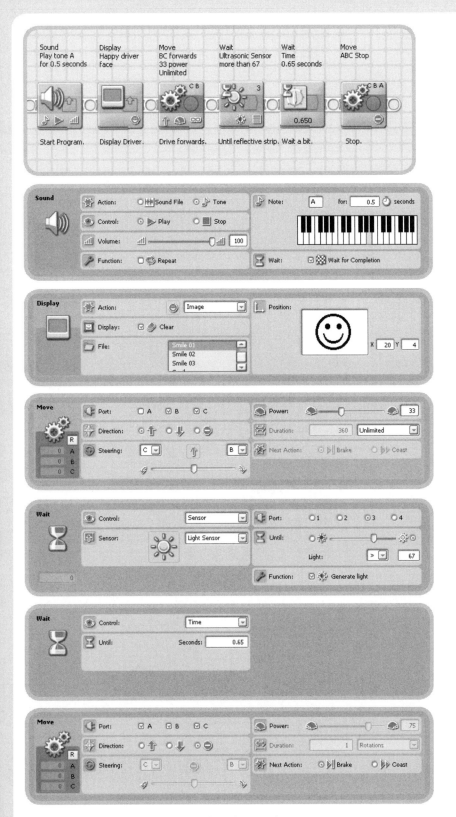

Figure 5-12: Detect the foil and drive forward 14 cm (5.5 inches).

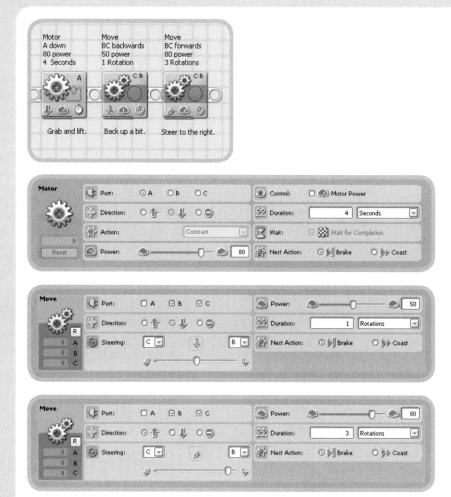

Figure 5-13: Close and raise the gripper, make a turn, and drive toward the bucket.

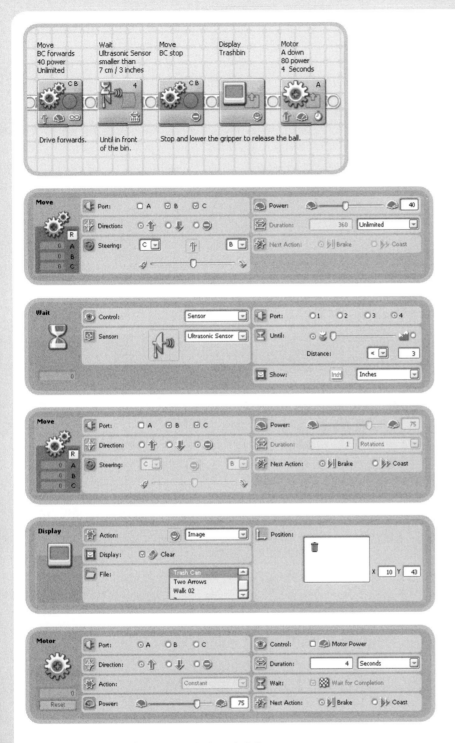

Figure 5-14: Stop in front of the bucket and release the ball.

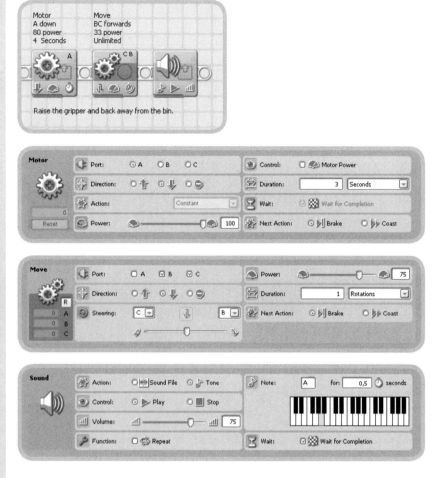

Figure 5-15: Raise the gripper and back away from the bucket.

further exploration

A real skid-steer loader comes with a large variety of attachments. If you have additional pieces, you might build the following:

* A real scoop—for example, the large black scoop with added pneumatics
* A snow blower
* A drill
* A brick grabber
* A rotating broom
* A tilted dozer blade
* A lawn mower
* A pneumatic demolition drill
* A soda can grabber
* Several pallets and a little warehouse to use with the forks
* Tracks instead of wheels
* A trailer to attach to the rear tow hook
* An ultra-large ball gripper for competitive ball grabbing

With a second NXT Brick or a compatible (Bluetooth-equipped) mobile phone, you can also add a joystick for remote control of the BobBot.

The program in this chapter is simple, but you can program the BobBot to do much more. Here are some suggestions:

* Have BobBot beep when backing up.
* Use ideas from the GrabBot's program (Chapter 8) to locate the ball and grab it.
* Change the smiley face displayed on the NXT Brick screen when BobBot is lifting and backing up.
* Add some custom sounds.

Figure 5-16: BobBot with changed arms and an additional scoop

6

RoboLock: a security system for your robots

Have you ever used an ATM? If you have, you know that you have to insert an ATM card and enter a password before getting money from it. ATMs have security measures that only allow certain people to get money from them. RoboLock imitates this process. It's designed to be a lock for your robot, allowing access only to cardholders with the correct code.

RoboLock works by reading a color-coded card that is inserted by a user. As the card is fed into the robot, a light sensor reads the code and determines if it's correct. If the code is incorrect, the robot ejects the card and tells the user to try again. If it's correct, the user is given access to the features of the robot secured by RoboLock. For the purposes of this book, we use RoboLock with a robot called Sentry. Sentry acts somewhat like a security camera, watching for and recording objects that come near the robot. When a user logs in to RoboLock using the correct card, he or she is given access to a menu that displays the number of objects detected by Sentry. As a bonus, the menu also displays the number of attempts made to log in with a false card—intruders beware!

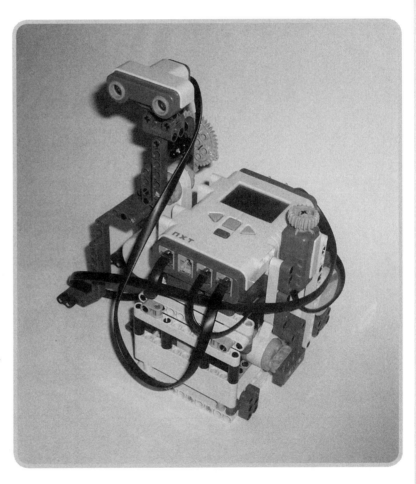

Figure 6-1: RoboLock with Sentry

building RoboLock

RoboLock is not a big robot, and it isn't complicated. Because it's meant to be combined with other robots, it doesn't use many valuable parts. For example, it uses only one motor,

Light Sensor, and Touch Sensor for electric parts (besides the NXT Brick, of course). In fact, the Touch Sensor isn't necessary, and it can be substituted for an NXT button if needed.

There are also instructions for building Sentry, which attaches to RoboLock like a module. Note that Sentry is not part of RoboLock itself, but it can be combined with it to make a secure guard robot. We offer it as an example of a robot that RoboLock can lock.

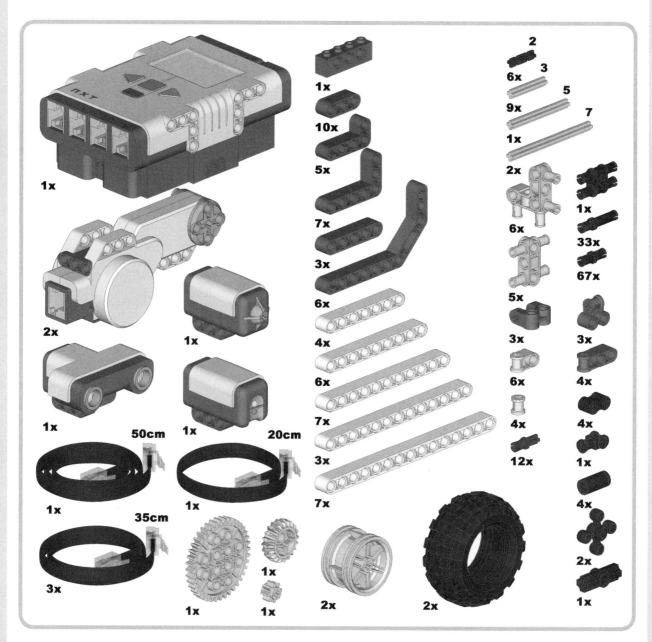

Figure 6-2: Bill of Materials

the light sensor module

1

1x   2x
1x   3x

2

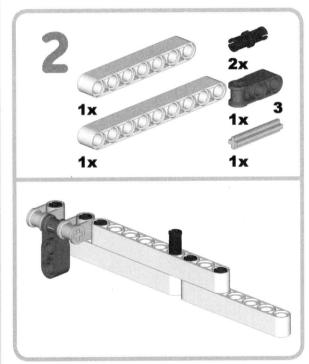

1x   2x
     1x   3
1x   1x

3

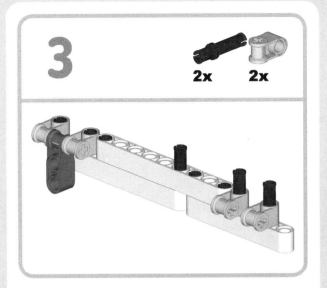

2x   2x

4

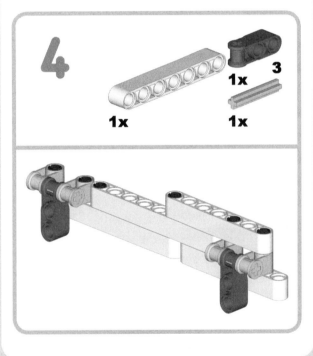

       3
1x
1x   1x

5

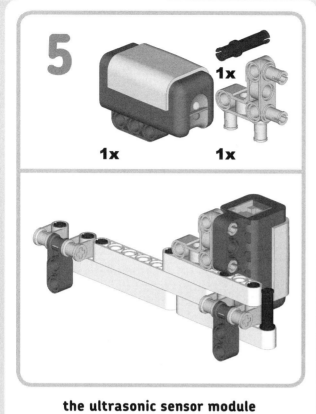

1x   **1x**
1x   **1x**

the ultrasonic sensor module

1

1x
2x

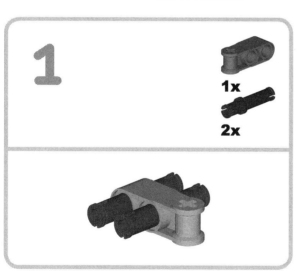

2

2x

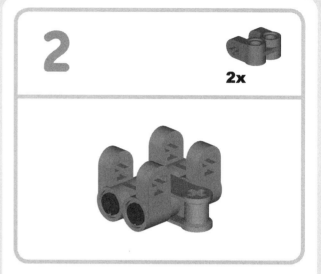

3

2
1x
1x
2x

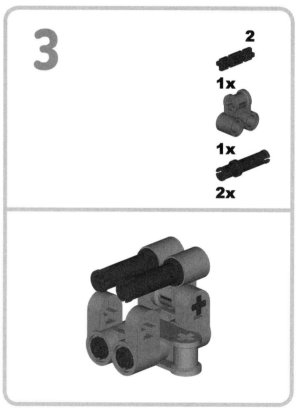

4

2

1x

1x   1x

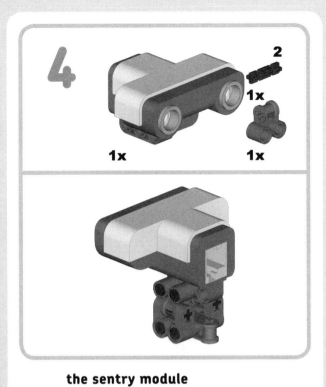

the sentry module

2

2x

2x

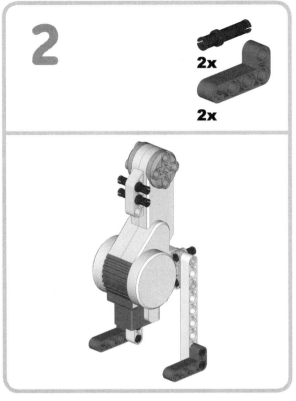

1

1x

8x

2x   2x

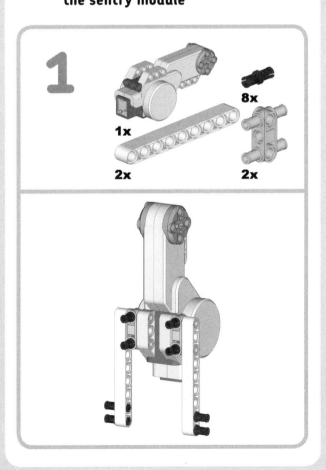

3

2x

2x

2x   2x

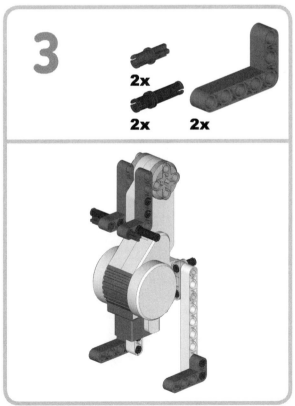

4

1x 1x 1x 7

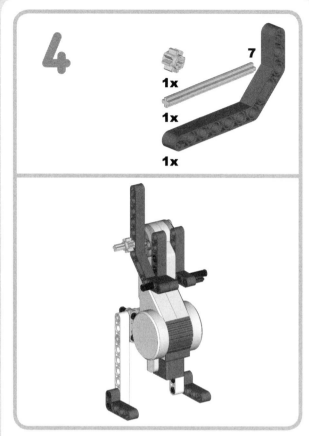

6

2 1x 3 1x 1x

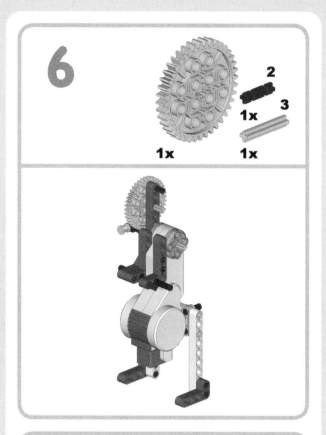

5

1x 3 1x 2x

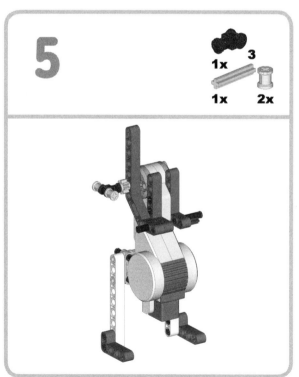

7

3 1x 2x 1x

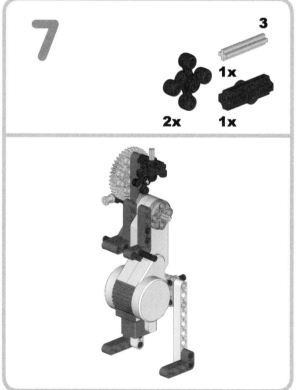

8

2

1x

1x

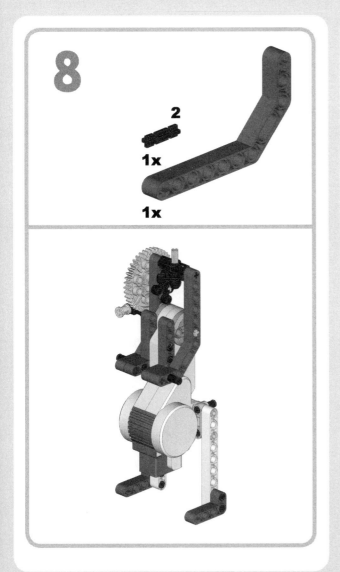

9

1x

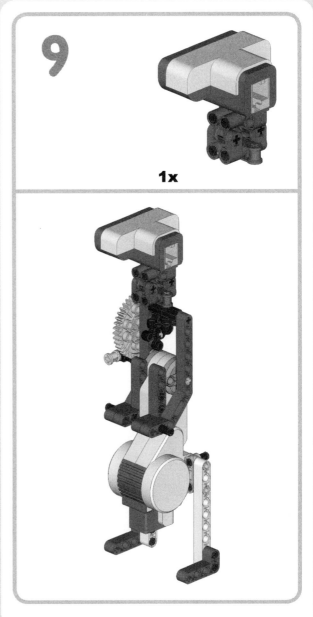

10

4x

1x

2x

7

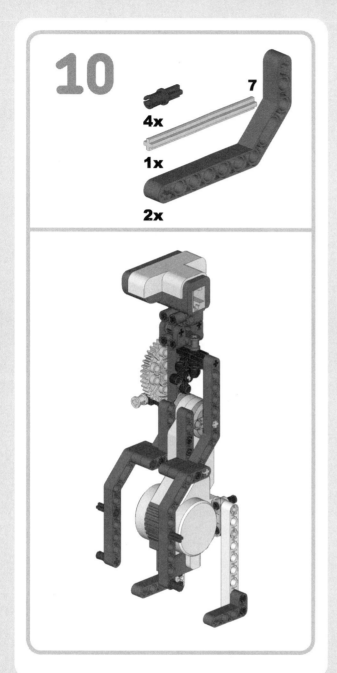

11

4x

3

4x

2x

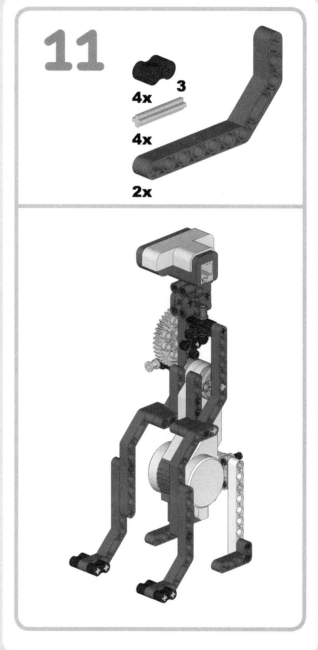

the front wall

1

2

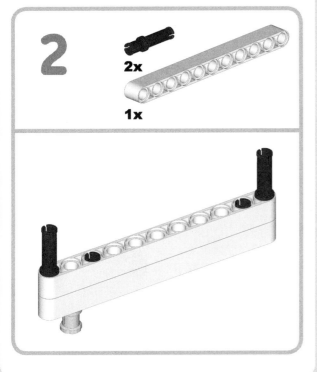

3

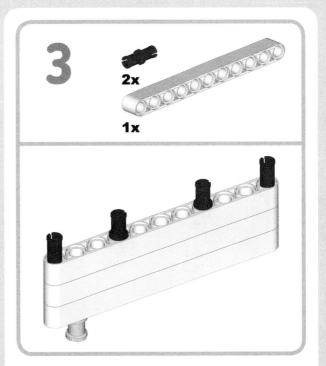

4

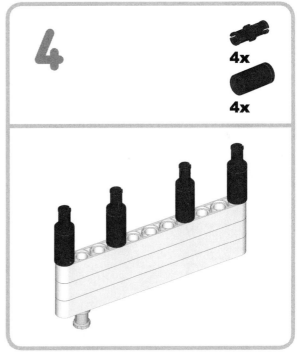

5

2x
1x
2x
2x

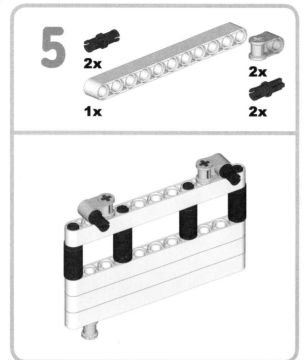

7

1x
1x
2x

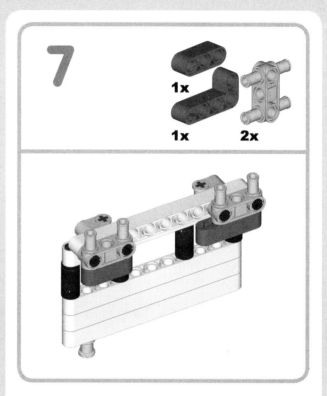

6

4x
1x

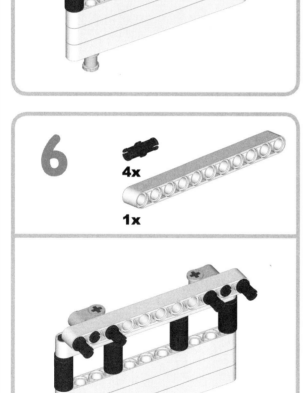

8

2x
1x

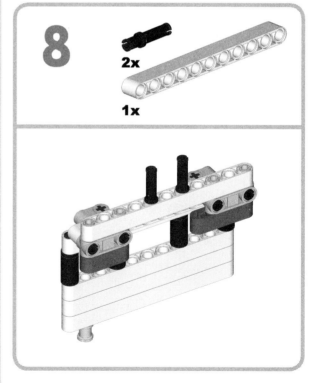

RoboLock

1

3x

1x

5x   2x

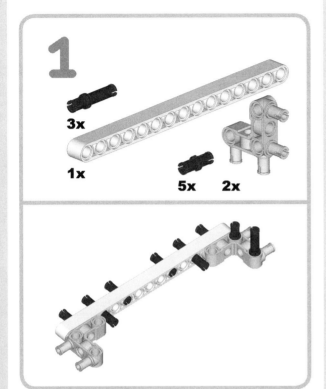

2

1x

2x   1x   5x

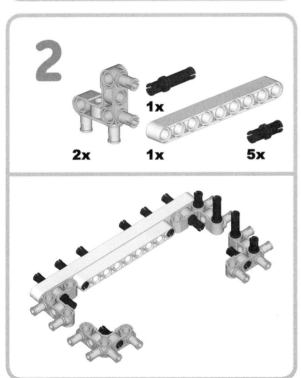

3

1x

1x

2x

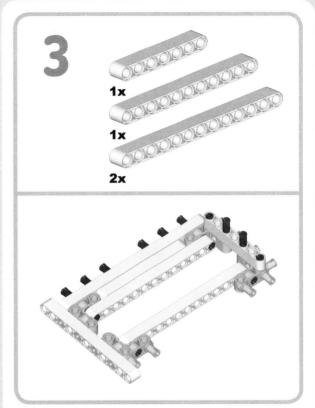

4

1x

1x

6x

1x

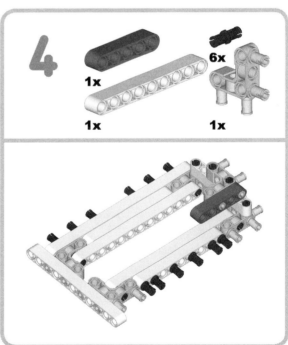

NOTE The beam in the preceding figure isn't supposed to be attached to the rest of the design yet. We'll do that later.

5

2x

4x

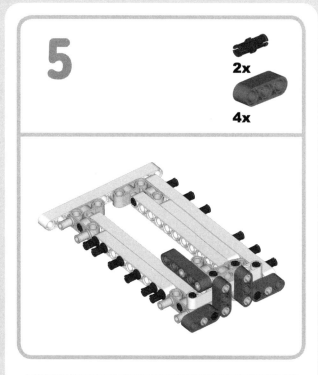

6

1x **1x** **1x**

1x **1x**

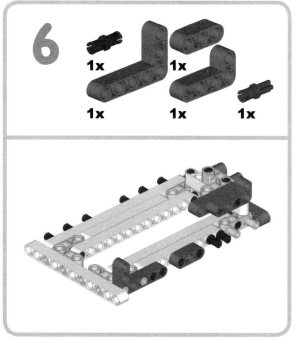

7

1x

1x

1x

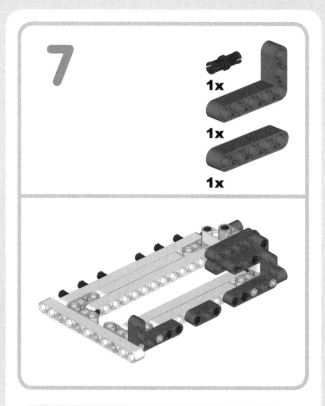

8

3x

2x

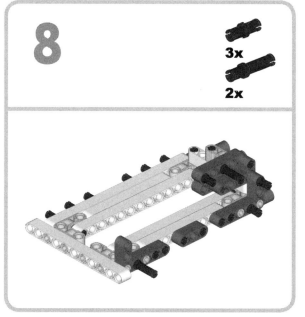

9

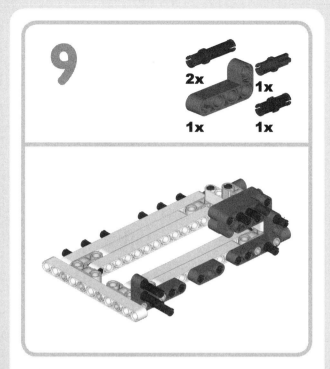

2x
1x
1x
1x

10

2x
2x

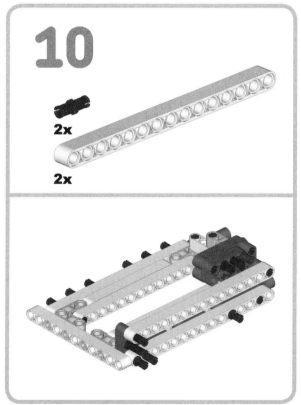

11

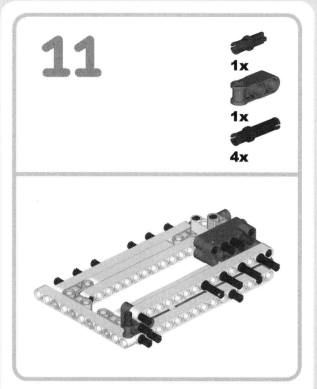

1x
1x
4x

12

1x
1x
1x

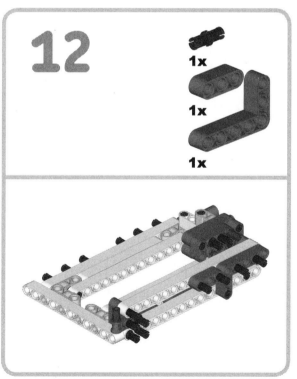

13

2x

1x

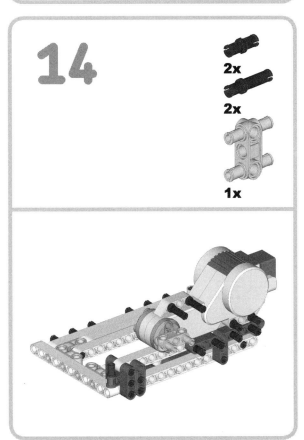

14

2x

2x

1x

15

2

1x 1x

1x 1x

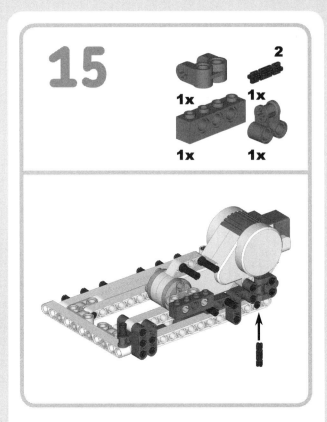

16

1x 5

1x

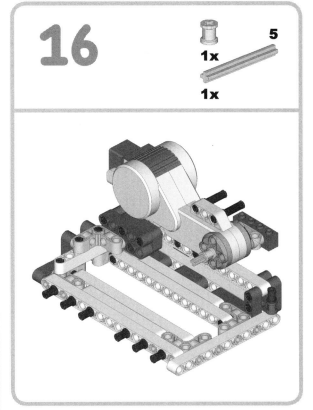

17

1x

1x

18

35cm

1x

1x

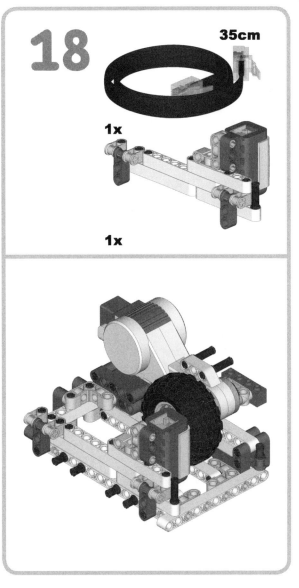

Plug one end of the cable into the Light Sensor, and leave the other end free outside the robot.

19

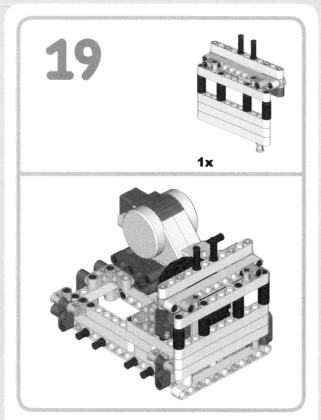

1x

20

6x

1x

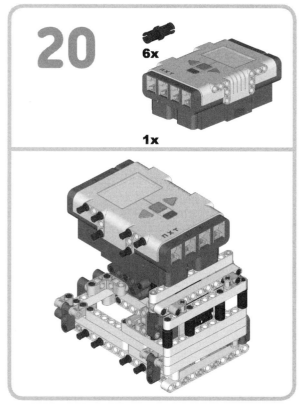

21

4x

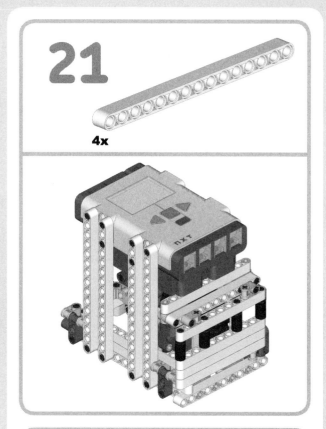

22

1x

3x

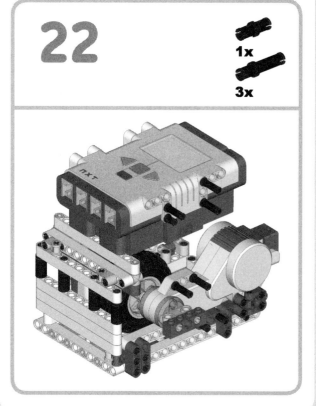

23

3x

1x

1x

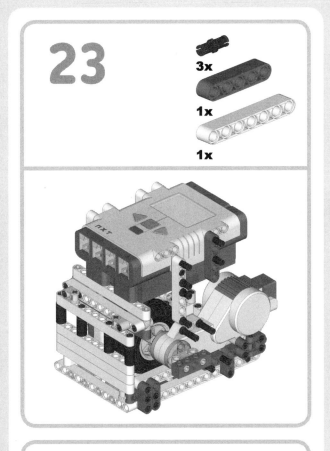

24

1x

1x

1x

1x

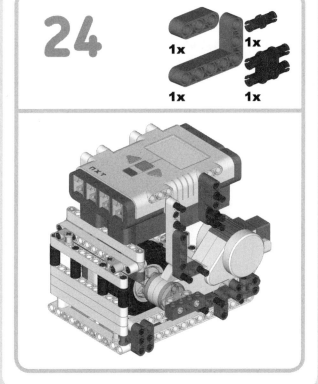

25

2

1x

1x

1x

1x

1x

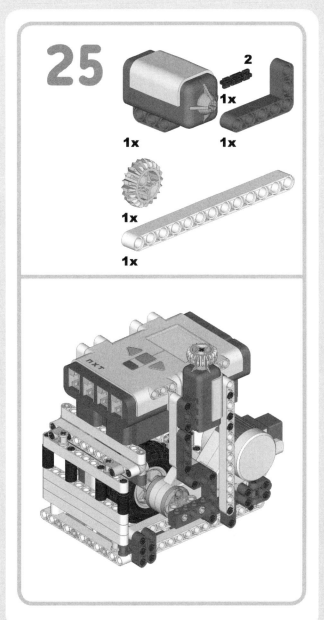

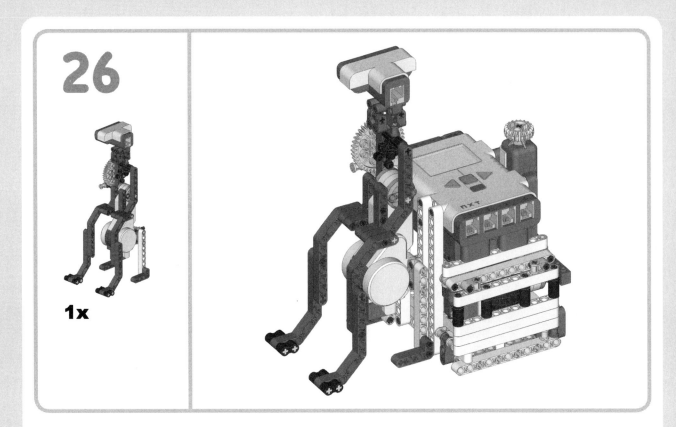

26

1x

connecting the cables

Now you're ready to connect the cables to RoboLock. The cable you connected to the Light Sensor in step 18 should be connected to input port 3. A 20-cm/8-inch cable should connect the card-feeding motor to output port B. A 50-cm/18-inch cable should connect the Sentry motor to output port C. Finally, two 35-cm/14-inch cables should connect the Touch Sensor and Ultrasonic Sensor to input port 1 and input port 4, respectively. Make sure none of the cables interfere with the robot's movements. The Ultrasonic Sensor cable should hang loosely and not be wrapped around any part of the robot. You can get an idea of how to arrange the wires from Figure 6-1 on page 199.

getting a secure card

The cards used with RoboLock aren't made with LEGO pieces but with thick paper. Figures 6-3 and 6-4 are two cards for RoboLock.

Photocopy these cards, cut them out, and tape them tightly to heavy paper (such as card stock paper), and you'll be ready to go. On the bottom of each card, you'll see a strip of black and white color squares. This is the code strip. As a bonus, the cards also include a place to sign your name. Notice that one card says *Secure Card* while the other says *False Card*. The Secure Card has the correct code, while the False Card has a (try to guess) false code. This card is included just for fun, so you can make sure RoboLock works correctly. To create your own cards, copy one of the cards in this book and modify the code strip.

The first three code squares on each card (the leftmost three) are colored black, black, and white, respectively. These squares make up the *synchronization strip*. They aren't part of the code strip but rather help RoboLock line the card up correctly. If these are changed, RoboLock may not work properly. The remaining six squares make up the *code strip*. This is the card's code, which is read by RoboLock's Light Sensor.

Figure 6-3: The secure card for RoboLock

Figure 6-4: A false card

programming RoboLock

In this section, we'll give you instructions for programming RoboLock to work with Sentry. However, you can modify the program to make RoboLock work with any other robot. Let's start with a quick overview of how RoboLock's program works.

When the program is started, RoboLock waits for a card to be inserted. When RoboLock detects a card, it feeds it through the reader using a motor. First it aligns the card correctly (in case the user inserts the card too quickly or too slowly) then feeds the card in at a constant rate. The Light Sensor looks at each code square, determines what color it is, and records it in a file as *w* for white or *b* for black. For example, if the code strip was arranged as white-black-white-black-white-white, the file would end up containing *wbwbww*. When all the code squares have been read, the program simply compares the recorded file with the correct code. If these do not match, the card is ejected with a "Try Again!", and the attempt is logged. If these do match, access is given to the menu. The menu greets the user and lists the number of UMOs and Intruders detected. *UMOs* stands for *Unidentified Moving Objects*, and it refers to objects detected by Sentry. *Intruders* refers to failed attempts to login.

While the menu is displayed, the user can turn Sentry on or off by pressing the orange NXT Enter button. RoboLock confirms the choice by saying "Start!" or "Stop!" Pressing the Touch Sensor *logs out*—that is, it causes RoboLock to eject the user's card and (if Sentry is turned on) start Sentry. Then RoboLock waits for someone to log in again, whereupon it stops Sentry and displays the menu again.

my block #1—code reader

We'll use two My Blocks in the program. The first is called *Code Reader*. It writes the *b*s and *w*s in a file to record a card's code. The program first puts the file's content into a text variable (CardCode) for storage, then it closes and deletes the file. Deleting the file allows it to create a new file (with the same name) to update the content, which is exactly what it does next. When this is done, it closes the file so it is ready to read the next file.

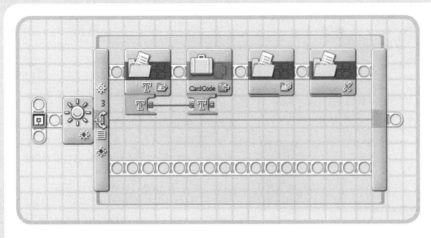

Figure 6-5: If the Light Sensor detects a white color, these blocks will store the Card Validation file's content in the CardCode variable and close and delete the file.

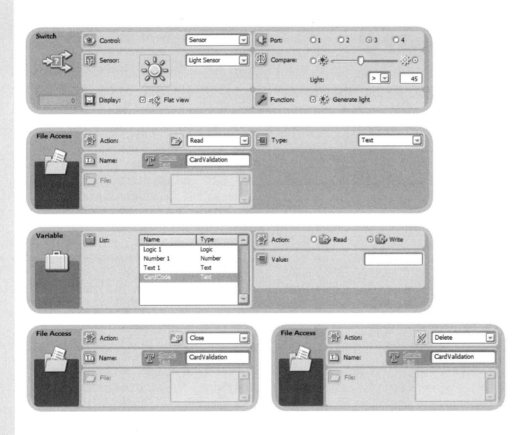

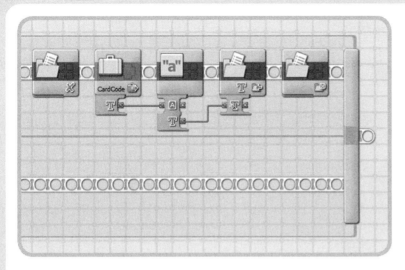

Figure 6-6: These blocks write the text in the CardCode variable along with a w to a new Card Validation file and close the file.

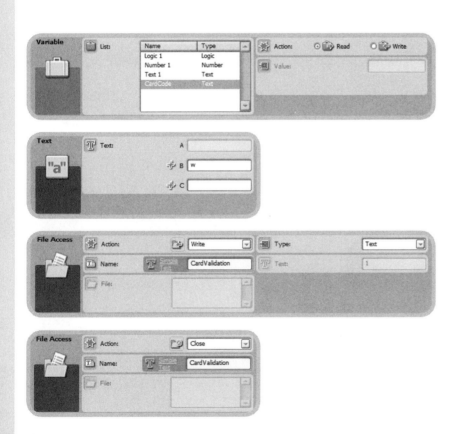

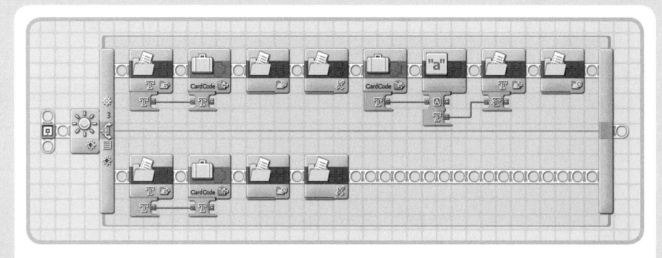

Figure 6-7: If the Light Sensor detects a black color, these blocks store the file's content in a variable and close and delete the file.

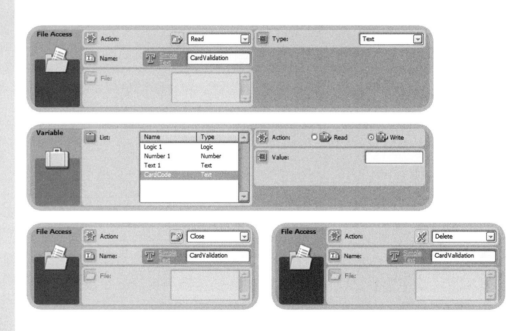

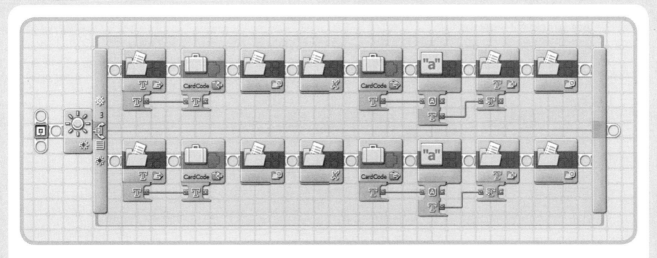

Figure 6-8: These blocks write the text in the CardCode variable along with a b to a new Card Validation file and close the file.

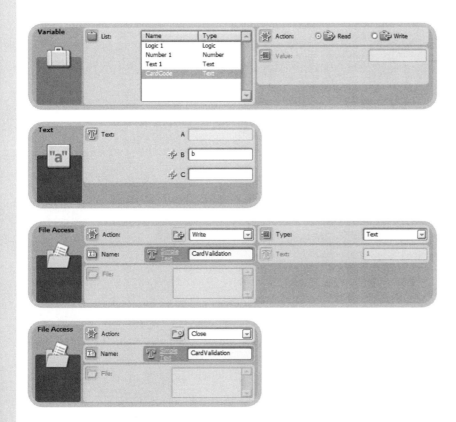

Create a My Block with these blocks and call it *Card Reader*. When you do this, you'll be ready to make the next My Block.

my block #2—menu

This My Block takes care of RoboLock's menu. When the correct card is inserted, this My Block will be started. It greets the user and displays the number of Intruders and UMOs.

If the user presses the NXT Enter button, it turns Sentry on or off by changing the value of a logic variable. It gives confirmation of the user's decision by making the robot say "Start!" or "Stop!" When the Touch Sensor is pressed, the My Block ends.

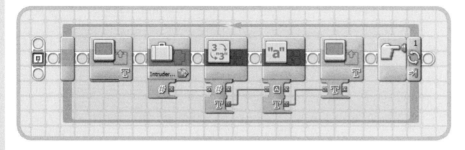

Figure 6-9: All the blocks in this My Block will be placed in the Touch Sensor Loop block that is shown. This makes the My Block exit when the Touch Sensor is pressed. The other blocks in the figure display Welcome User *and the number of Intruders.*

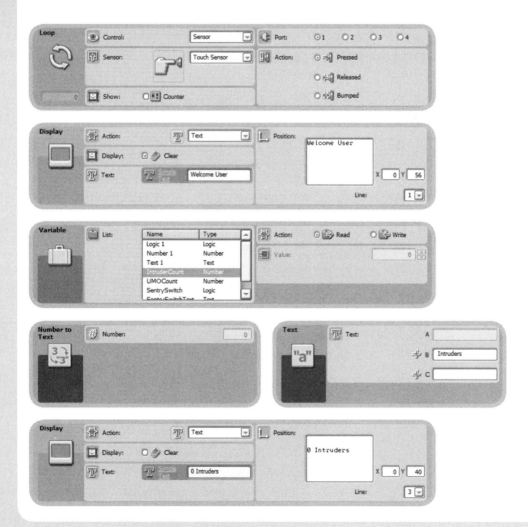

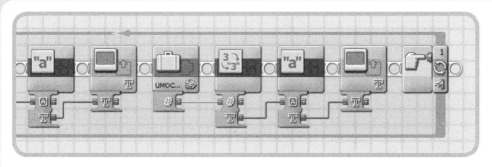

Figure 6-10: These blocks display the number of UMOs detected by Sentry.

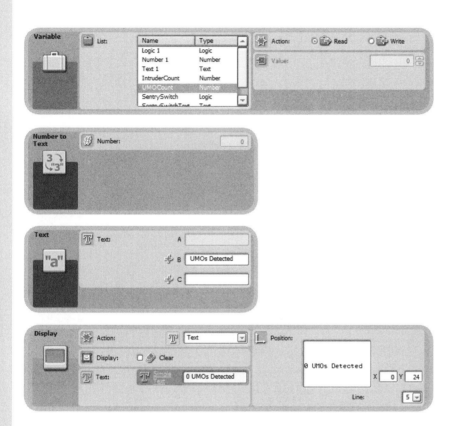

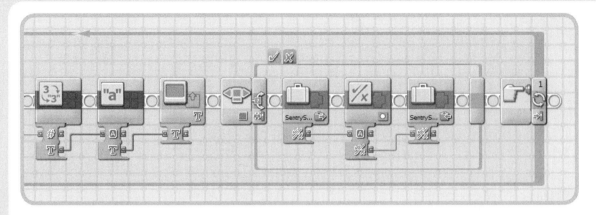

Figure 6-11: If the NXT Enter button is pressed, these blocks change the value of the logic variable SentrySwitch. This determines whether Sentry will be on or off, as we'll see later in the program.

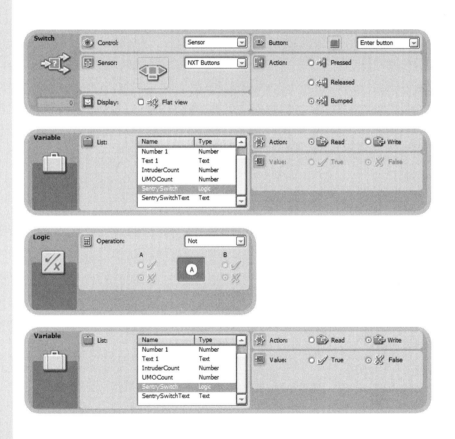

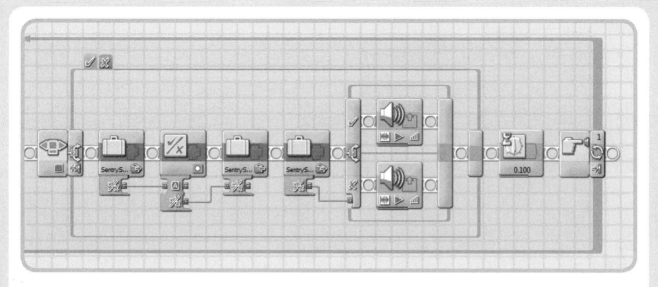

Figure 6-12: These blocks make RoboLock say "Start!" or "Stop!" depending on whether Sentry was turned on or off (which is determined by the SentrySwitch variable).

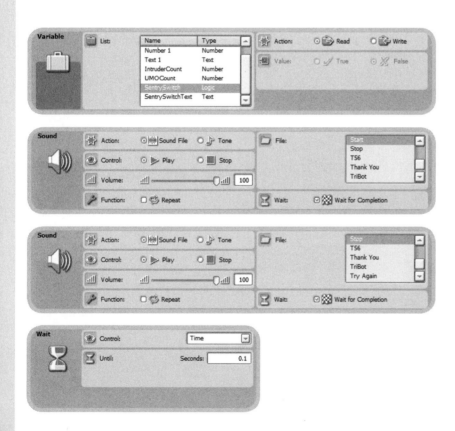

Once again, put all these blocks into a My Block called *Menu*. Now it's time to make the main program.

main program

This program uses the Code Reader and Menu My Blocks
and makes RoboLock work like it's supposed to.

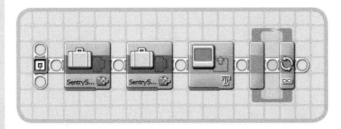

Figure 6-13: These blocks set the default values of three variables used later and display
<Insert Card>. The Loop block will be used to let a user log in and out as many times as he or
she wants by continually looping the card-insertion/menu-display/logout sequence.

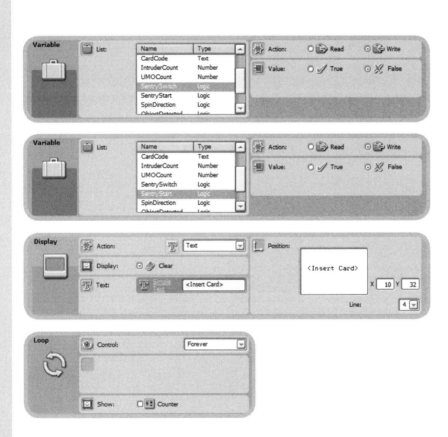

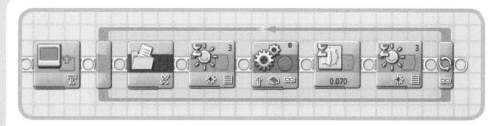

Figure 6-14: These blocks delete the Card Validation file (in case it was used in an earlier program) and wait until a black color is detected, which means a card has been inserted. Another Light Sensor Wait block waits until a white color is detected, which means the card is lined up.

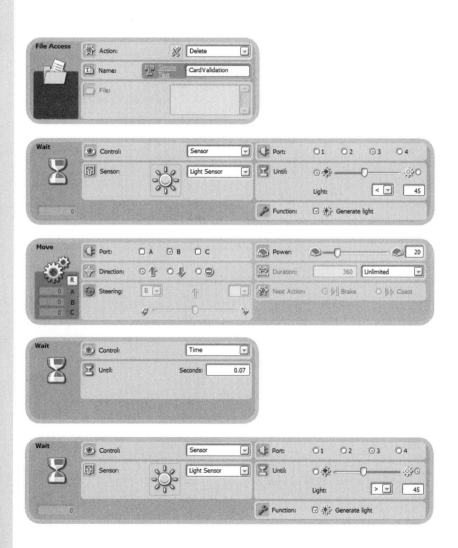

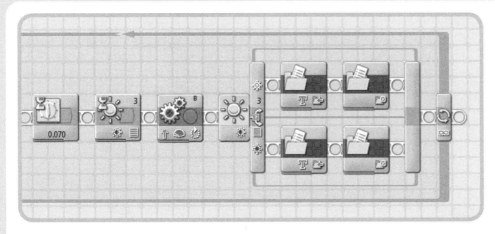

Figure 6-15: The Move block makes Motor B move the card such that the first code square is under the Light Sensor. Then the remaining blocks put a b or w in the Card Validation file, depending on the color of the code square.

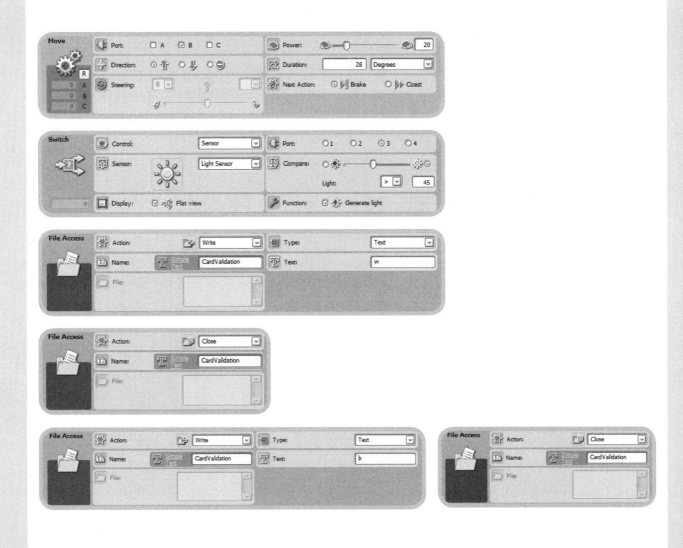

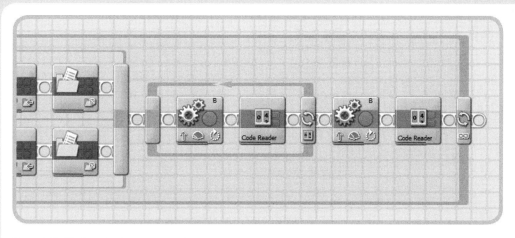

Figure 6-16: The first three blocks repeatedly make Motor B move the card to the next code square and record the color in the Card Validation file (using the Code Reader My Block). After all but one of the squares have been read, the Loop block exits, and the remaining two blocks read the last square.

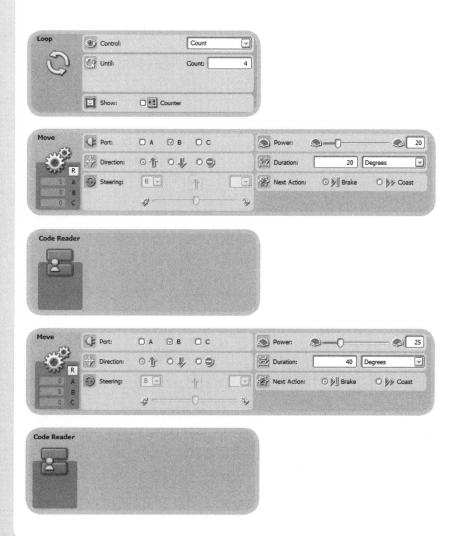

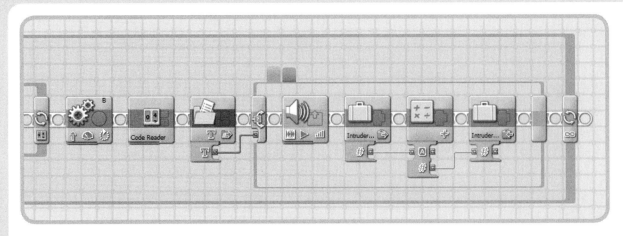

Figure 6-17: The Text Switch block compares the Card Validation file with the correct code. If they don't match, the other blocks make RoboLock say "Try Again!" and increment the IntruderCount variable to record the failed attempt. Note that you need to put these blocks in the 0 condition of the Switch block.

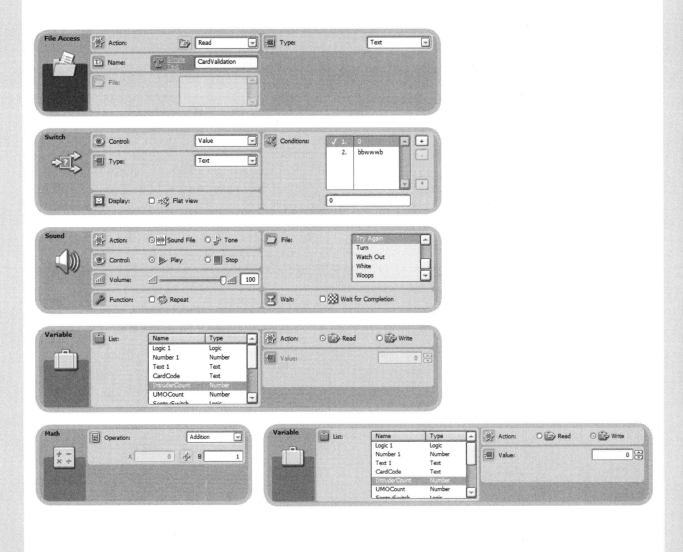

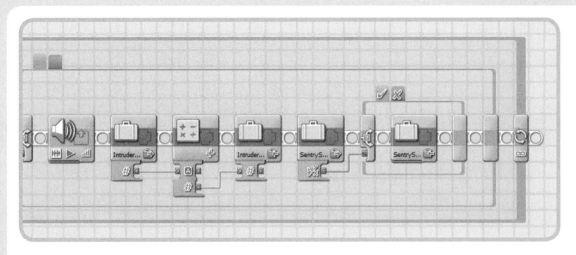

Figure 6-18: These blocks turn Sentry on if it's supposed to be on.

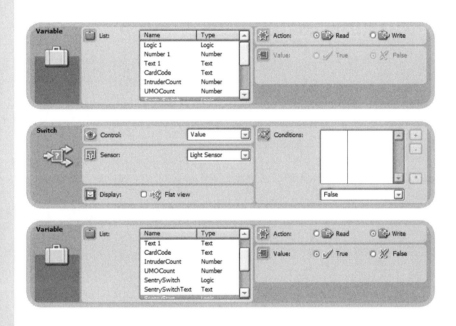

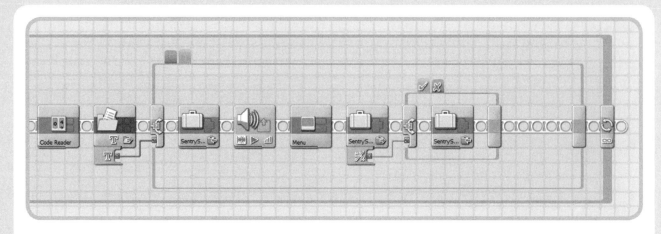

Figure 6-19: If the code was correct, these blocks turn Sentry off (by making the value of the SentryStart variable false), make RoboLock say "Hello!", and display the menu by using the Menu My Block. When the My Block ends, the remaining blocks turn Sentry on if it's supposed to be on. Note that these blocks should be put in the bbwwwb condition of the Switch block.

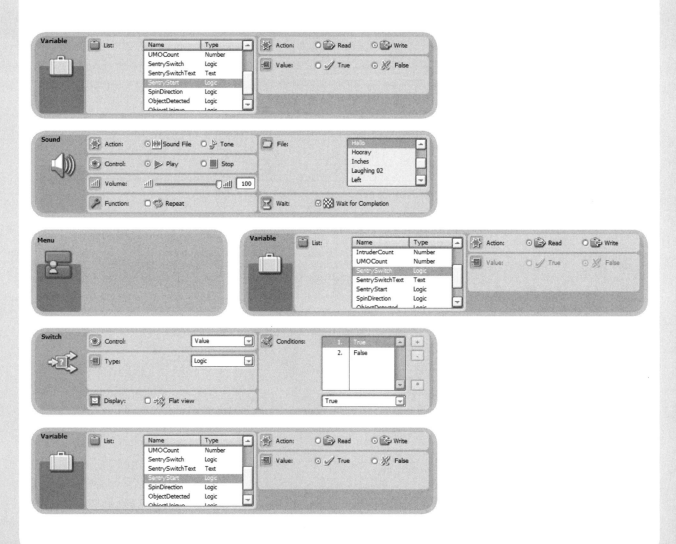

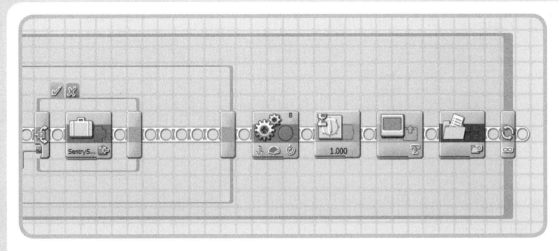

Figure 6-20: These blocks eject the card, wait a second before displaying <Insert Card> again, and close the Card Validation file.

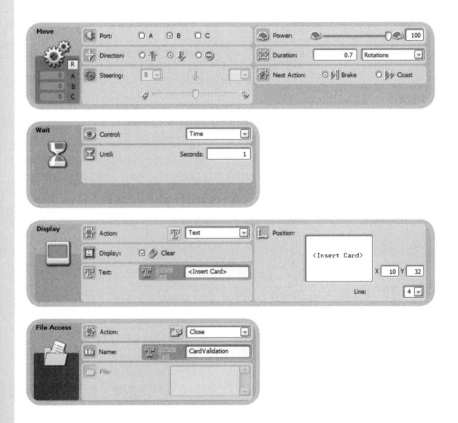

Now we'll program Sentry's tasks. This code will run at the same time as the previous code so that Sentry can be on while RoboLock is waiting for a card to be inserted.

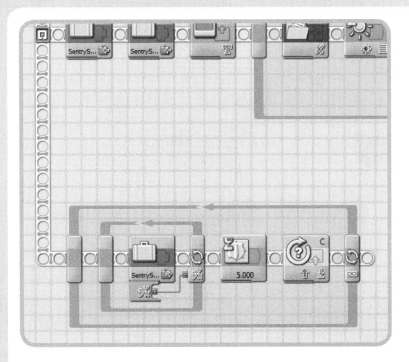

Figure 6-21: These blocks wait for Sentry to be turned on (by the value of the SentryStart variable being set to true), wait for five seconds to let the user get out of the way, and then reset motor C's rotations. The Loop Forever block will enable Sentry to be turned on and off as many times as desired.

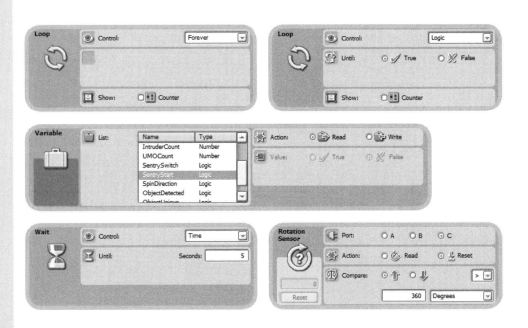

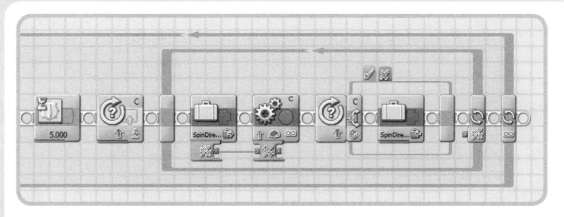

Figure 6-22: These blocks start motor C in the direction determined by the SpinDirection variable and test to see if motor C has turned more than 870 degrees in a forward direction. If it has, the blocks change the value of SpinDirection to make motor C turn the other way. The Logic Loop block will keep Sentry on until it's turned off.

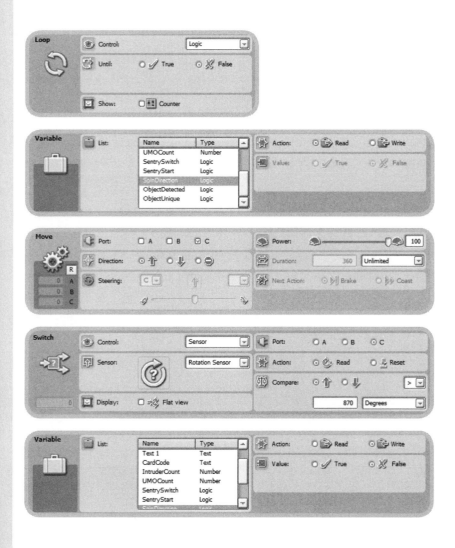

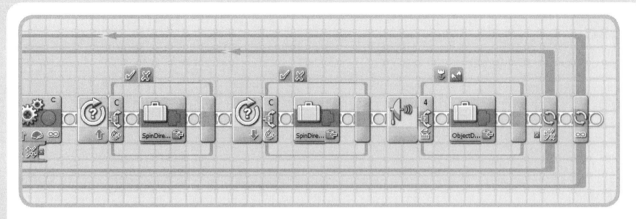

Figure 6-23: The first two blocks determine if motor C has turned more than 870 degrees in reverse, and they change the value of SpinDirection if it has. The next block determines if an object is in range of the Ultrasonic Sensor. If it is, the ObjectDetected variable is set to false. *This variable will be used later to help Sentry detect only unique objects.*

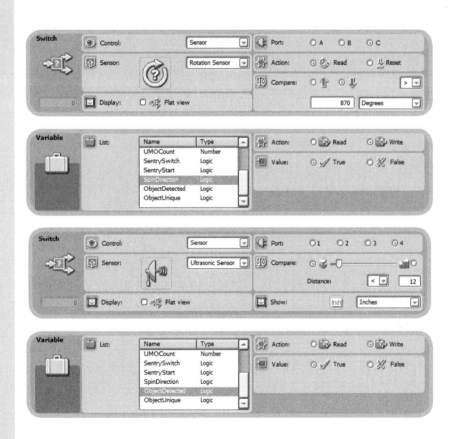

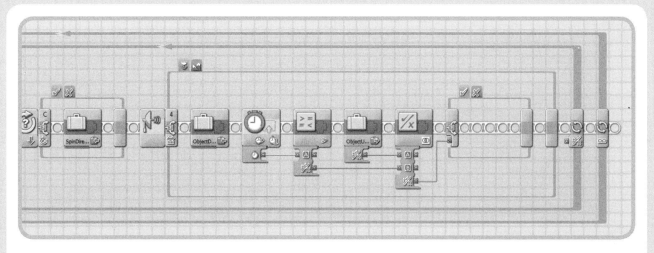

Figure 6-24: These blocks test to see if Timer 1 is greater than five seconds and the ObjectUnique variable's value is true. Timer 1 and the ObjectUnique variable are used to help Sentry only detect unique objects. Timer 1 is used to wait for five seconds after seeing an object before recording any more UMOs. The ObjectUnique variable is used to wait until the object leaves.

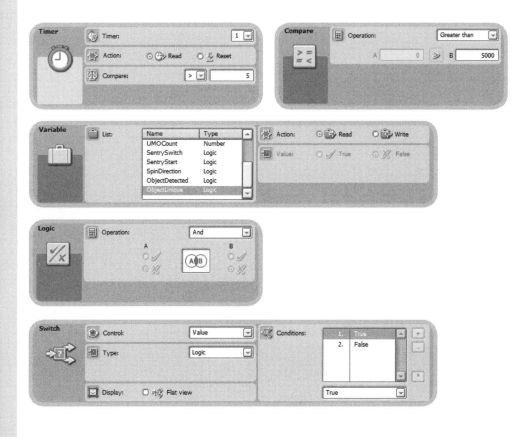

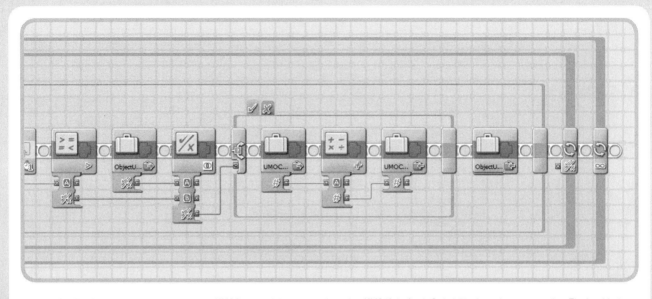

Figure 6-25: The first three blocks increment the UMOCount variable to record another UMO if the Logic Switch block receives a true value. The last block sets the ObjectUnique variable's value to false. This indicates that the current object has already been recorded.

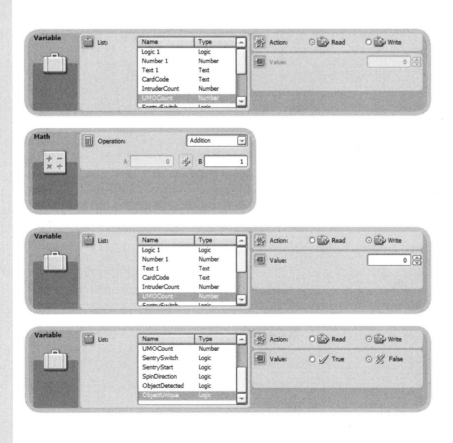

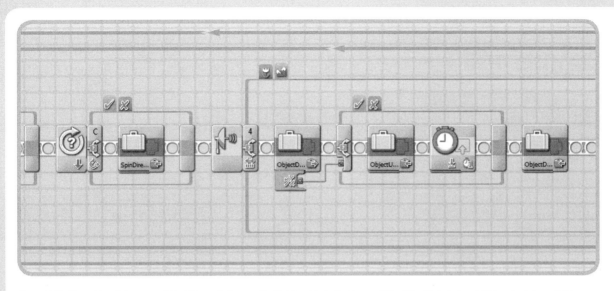

Figure 6-26: If no object is in range of the Ultrasonic Sensor, the first two blocks test to see if the ObjectDetected variable's value is true. If it is, an object has just gone out of range, so the ObjectUnique variable's value is set to true and Timer 1 is reset. Then the last block sets the ObjectDetected variable to false to indicate that an object hasn't just gone out of range.

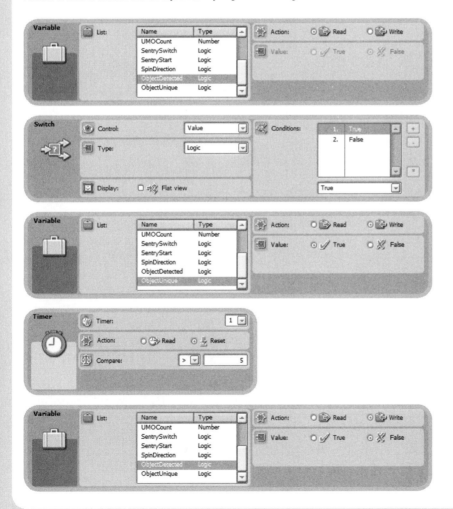

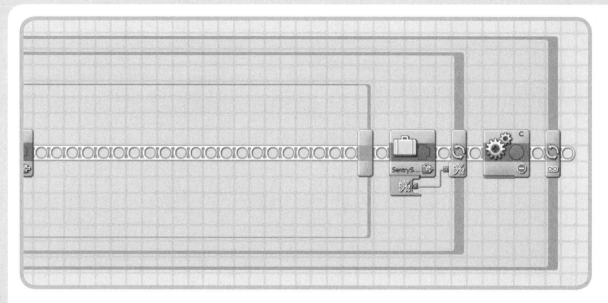

Figure 6-27: The value of the SentryStart variable is wired into the Logic Loop block placed earlier to make it exit when Sentry is turned off. When this happens, the Move block turns motor C off to stop the Ultrasonic Sensor from turning around.

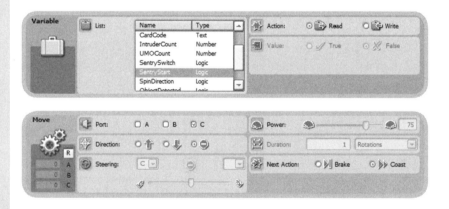

troubleshooting tips

* Always insert cards face up and with the code squares on the left side because this will enable the Light Sensor to read them. A card should be inserted at a slow, steady rate until the motor starts pulling it in. When you feel the motor pulling on the card, you should let go.

* Before turning Sentry on, you'll need to make sure the Ultrasonic Sensor is pointing directly toward the back of RoboLock, the opposite direction that is shown in Figure 6-1. You can do this by turning the handle connected to the motor that rotates the sensor.
* If RoboLock isn't detecting code squares correctly, you may need to adjust the values of the various Light Sensor Wait and Switch blocks.

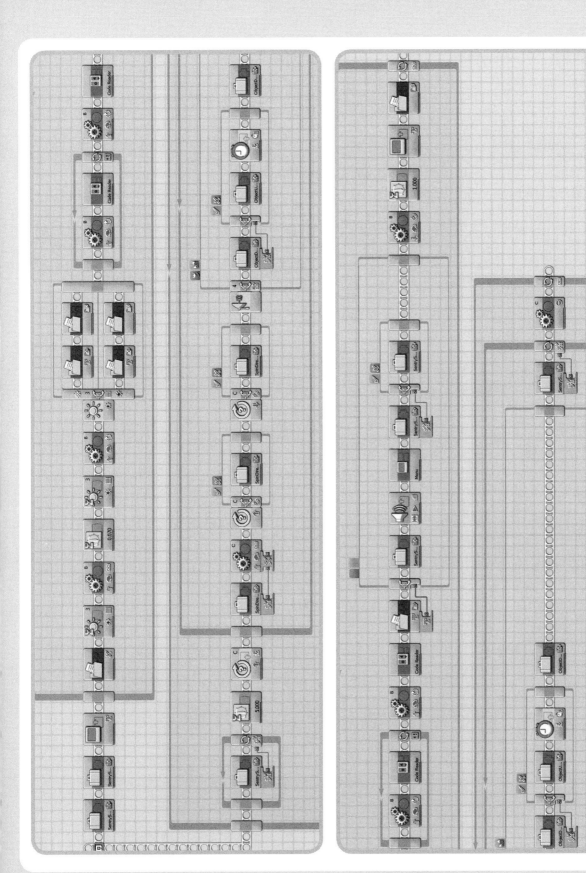

Figure 6-28: The final RoboLock program

the hand: a robot for those dirty jobs

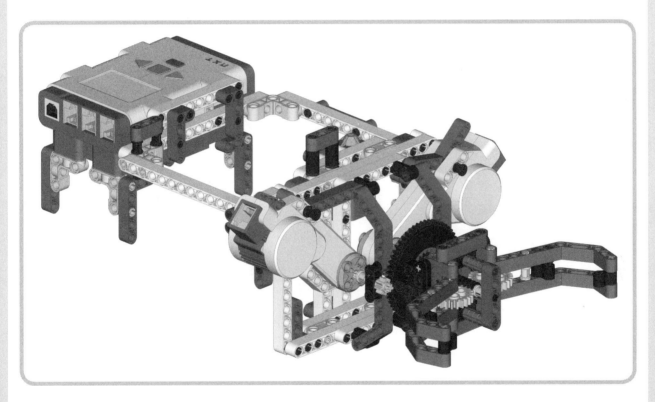

Figure 7-1: The Hand

Robotic hands come in many shapes and sizes, and they have many purposes. They have been used to do dangerous work (like handling chemicals or your brother's dirty gym socks). They've been used to do tasks that require extremely delicate movements (such as assisting with surgery), and they have been used to do boring, repetitive tasks (like riveting a door on a car during manufacturing).

Another application for a robotic hand is prosthetics. *Prosthetics* is the science of making and fitting artificial body parts. Suppose you lost your hand in a car accident. If you were outfitted with an artificial hand, that would be called a *prosthetic*.

The Hand that we are going to create here is not as complex as the human hand, but it does have some of the same basic movements. The Hand can close and open its fingers to grasp and release an item. It can also rotate 360 degrees at the wrist. (And it can keep revolving for as long as your batteries last.)

The Hand sits on your own arm, and it is operated via the buttons on the NXT Brick and a Touch Sensor. If you have larger or smaller hands, you might need to adjust the handle and Touch Sensor positions accordingly.

The fingers are wide and only come together at the tips. This means that The Hand can only grab things that are wider than the space between the closed fingers. See "Building Modifications for The Hand" on page 269 for ways to enhance The Hand and allow it to pick up smaller objects.

There is one issue to be aware of: If you hold the center axle and look closely when turning the wrist axle, you will see the fingers slowly open or close. This happens because a turntable allows a long axle to go through the inside of the gear and rotate the fingers on the other side of the turntable. You can also put a gear on the outside of the turntable and rotate the entire piece (the wrist movement). This allows the wrist to rotate 360 degrees. However, because the axle that runs through the middle of the turntable is locked (it's held in place by a motor and only rotates when the motor rotates), it means that as you turn the wrist, the fingers will open or close slightly. Figure 7-14 on page 266 shows how the NXT-G program solves this problem.

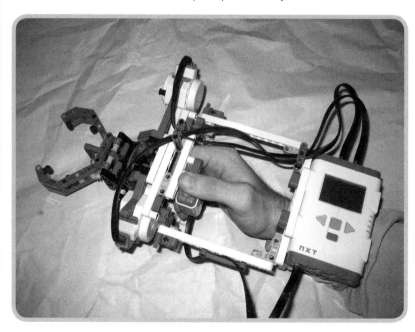

Figure 7-2: The Hand shown on the author's hand

building the hand

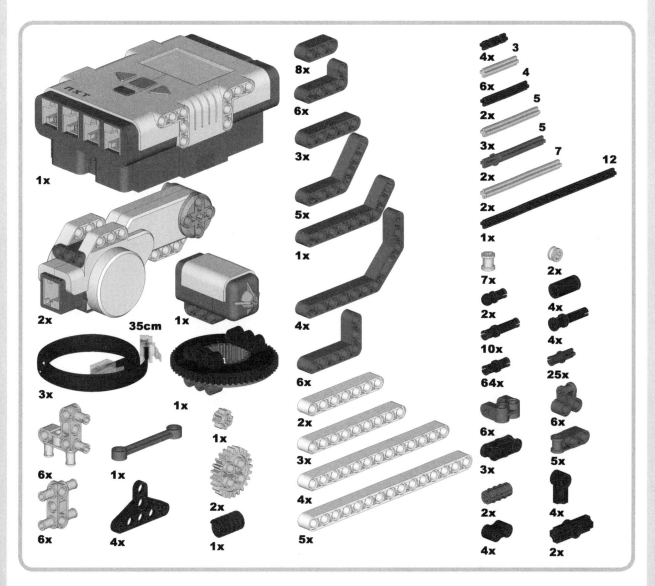

Figure 7-3: Bill of Materials

the NXT base

1

4x   4x

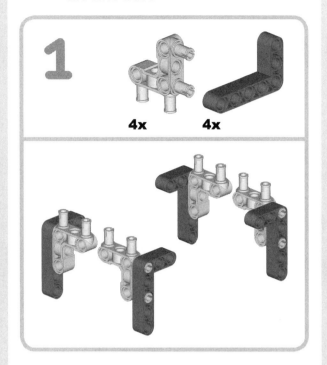

2

1x

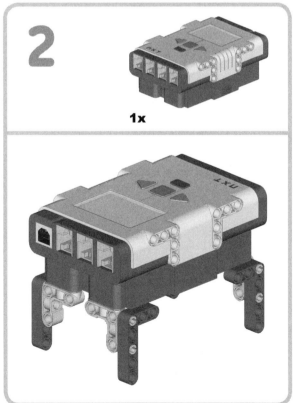

the main frame

2x
2x
1x

1

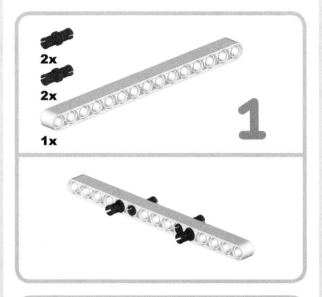

2

2x

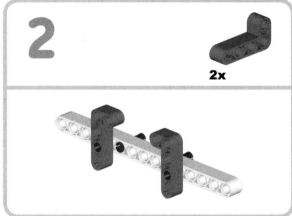

3

4x

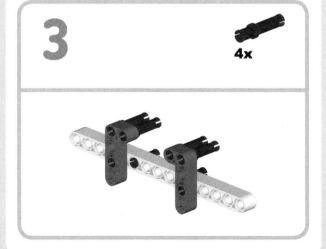

4

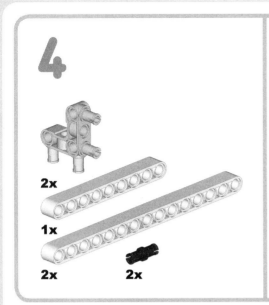

2x

1x

2x 2x

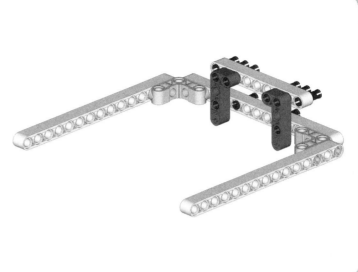

5

2x

2x

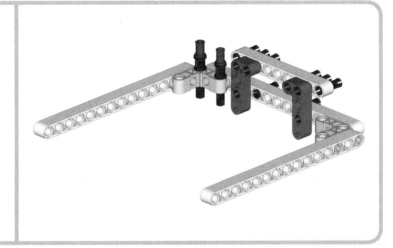

6

1x 2x

7

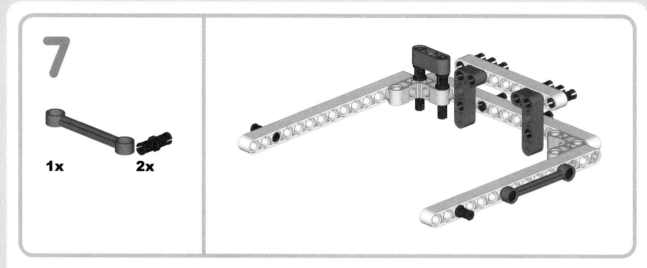

1x   2x

8

1x   1x

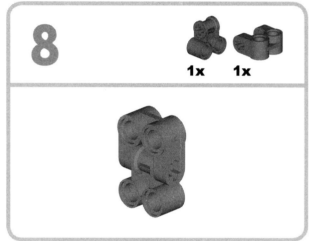

9

2

1x

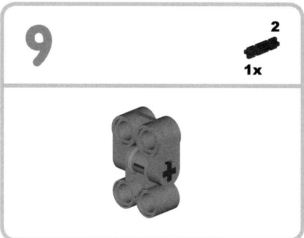

Make two of these.

10

2x

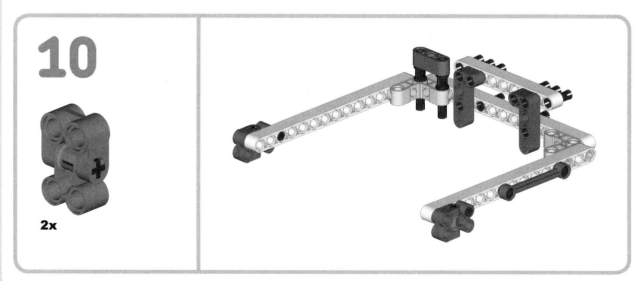

11

4x

12

4x

2x

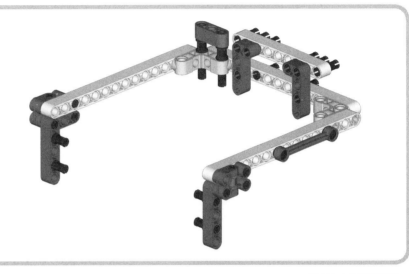

13

2x

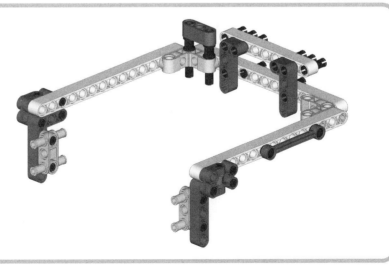

14

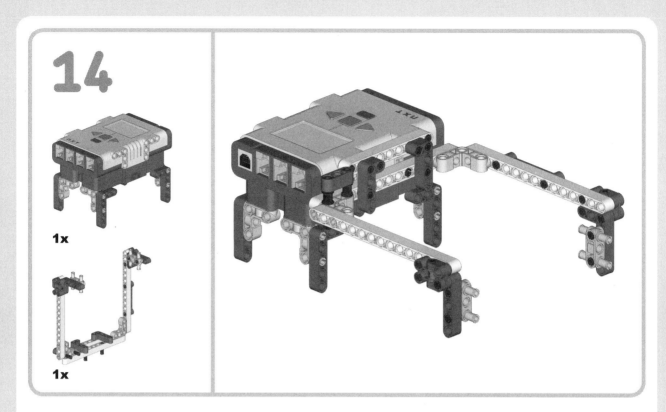

1x

1x

the handle

1

4x

2x

2

2x

2x

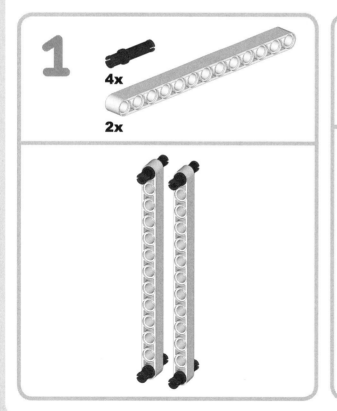

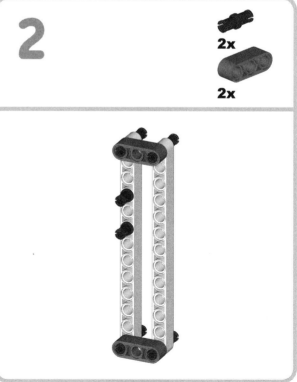

3

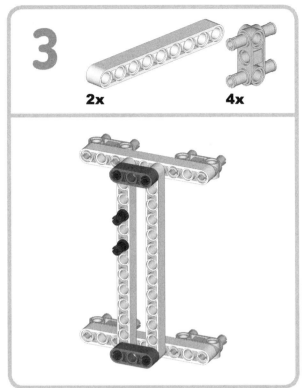

2x   4x

4

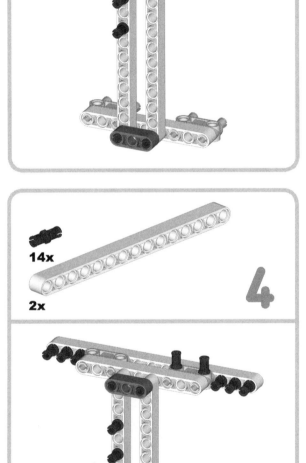

14x

2x

5

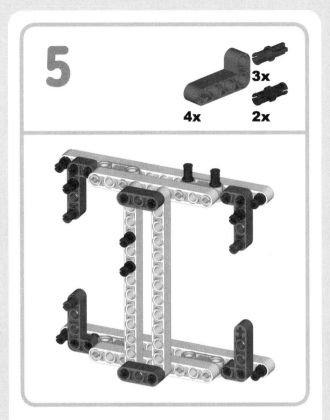

3x

4x   2x

6

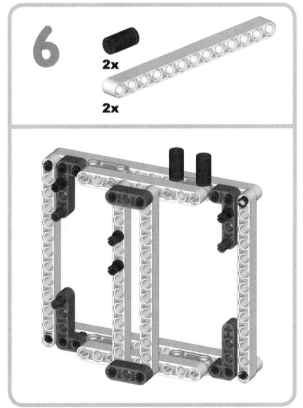

2x

2x

7

7x

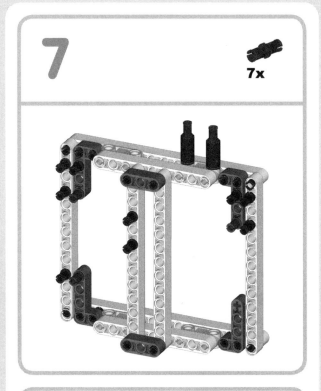

8

2x 2x

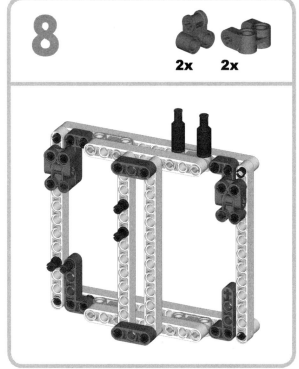

Notice there is no axle in the pieces added here. It will be added later when the parts are assembled into the final model.

9

2x

1x

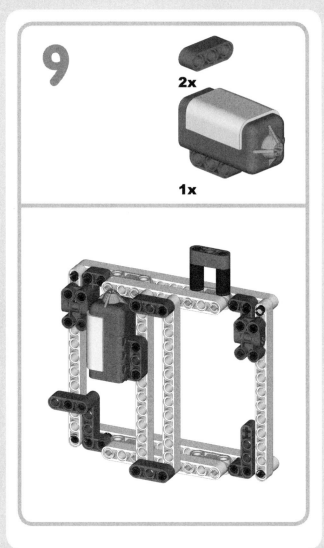

10

1x

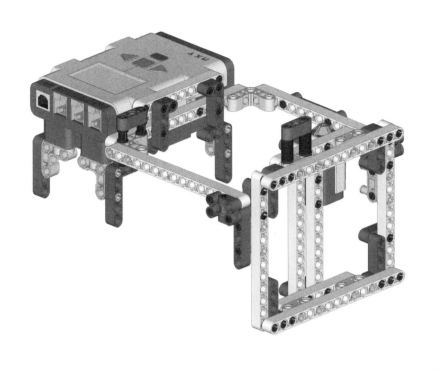

11

5.5

2x

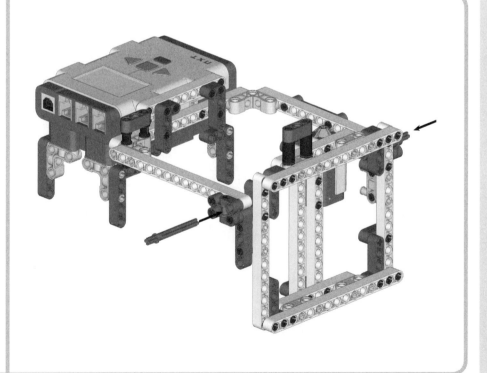

the hand

1

2x
1x

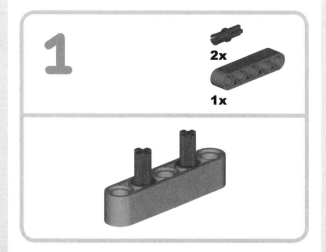

2

2x
1x

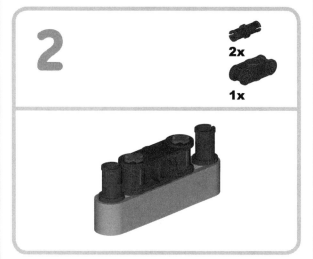

3

2x

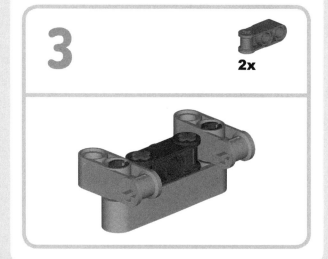

4

1x
1x   7
2x

5

1x   3
1x
1x   5
1x
1x

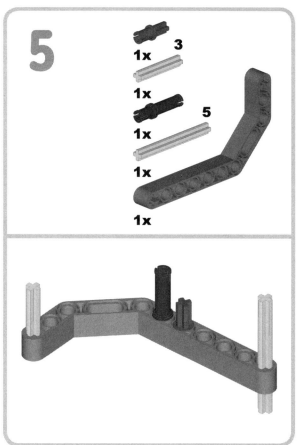

6

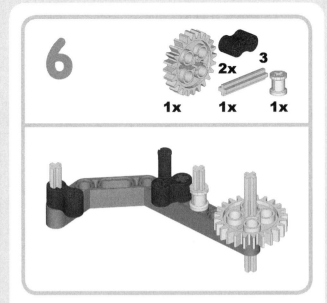

2x   **3**

1x   **1x**   **1x**

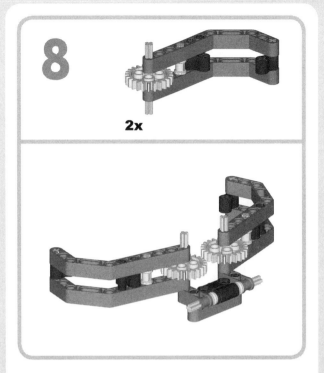

8

2x

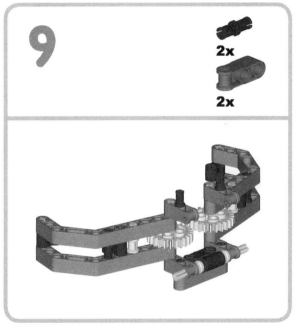

7

1x

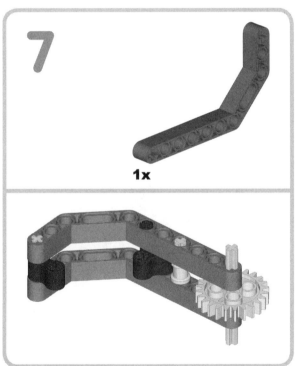

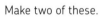

Make two of these.

9

2x

2x

10

2x

1x

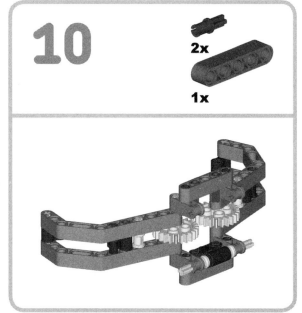

11

1x

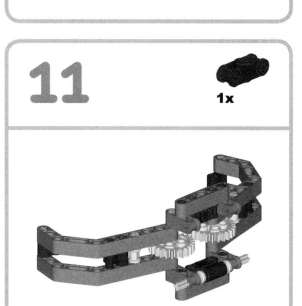

12

7

1x

1x

2x

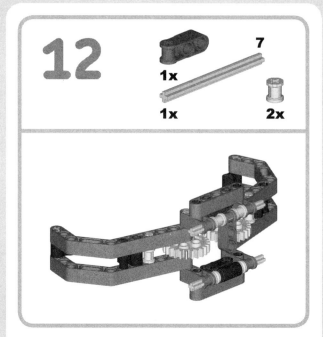

13

2x

14

2x
2x

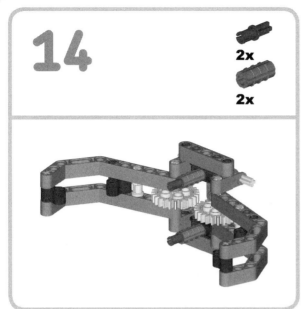

16

1x

15

12
1x
1x
1x

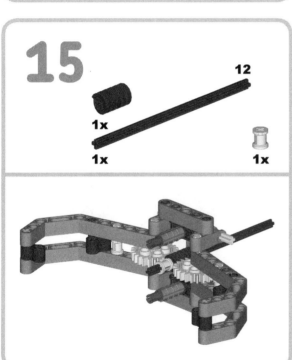

17

1x 1x

When you insert the axle with the worm gear on it, make sure the fingers are angled open as far as they can go. Insert the axle. The worm gear will engage the gray gears and bring the fingers forward.

18

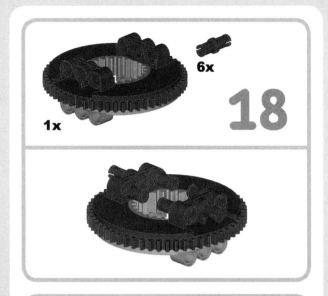

6x

1x

21

2x

19

1x   1x

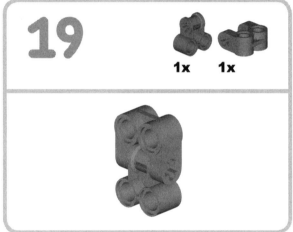

22

4x

20

2

1x

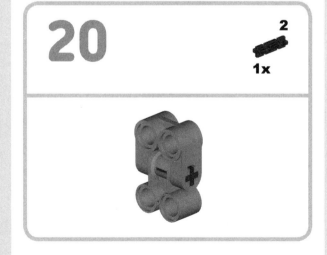

Make two of these.

23

2x

2x

24

1x

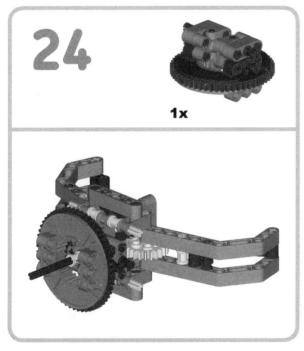

25

4x

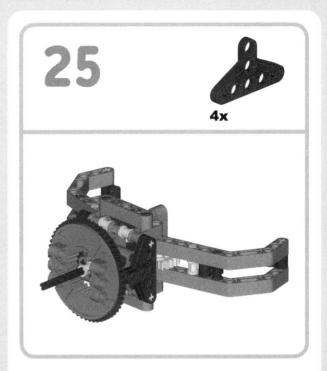

It may be easier to attach one triangle piece at a time on each side and then add the second triangle pieces.

26

2x **3**

2x

2x **4**

2x

2x

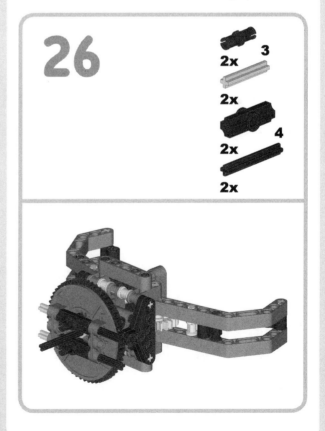

27

1x
4x

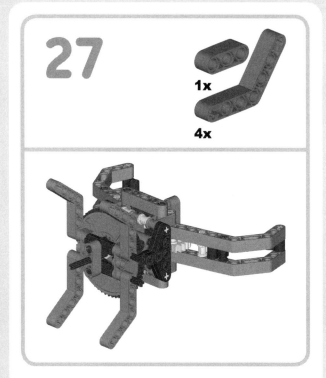

You can test the fingers and wrist by holding the beams and turning either the gray axle (to rotate the wrist) or the black axle (to open and close the fingers). If the wrist doesn't turn or the fingers don't work, go back and review the assembly. When you attach the motors, you can't manually rotate the axles anymore.

28

1x
1x
5
1x
1x

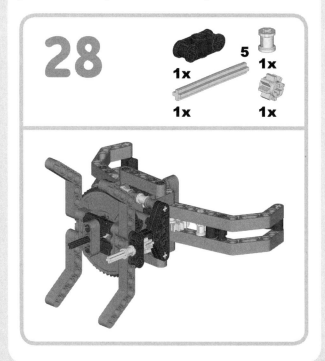

29

4x
8x

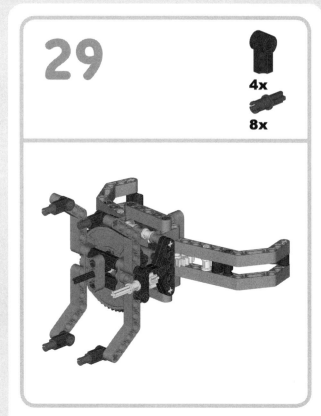

30

3x
1x

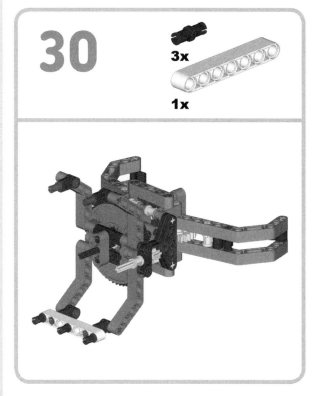

31

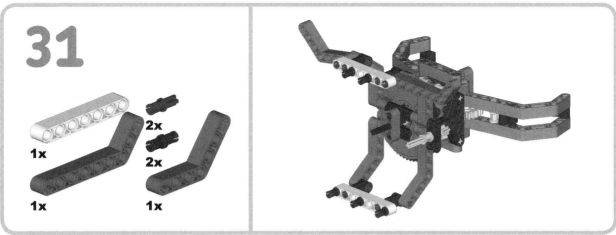

1x   2x   2x   1x   1x

32

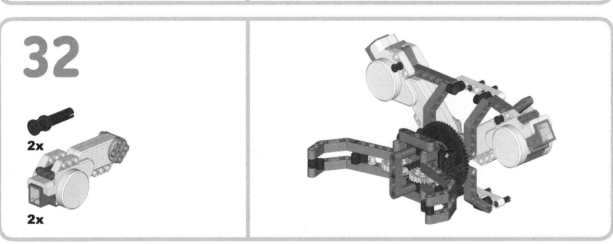

2x
2x

Note that the entire model has been rotated to the right. Be sure to insert the axles into the correct motor. The black axle on the left goes into one motor, and the gray axle on the right goes into the other motor. Also, inserting the long black pins can be tricky. You might need to wiggle the motor a little to get the pins and holes to line up properly.

33

1x

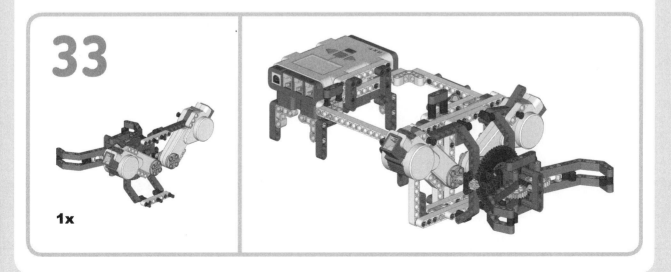

connecting the cables

The Touch Sensor connects to input port 1, the left motor that turns the center axle connects to output port C, and the right motor that turns the outside axle connects to output port B. Notice how the three wire guides hold the wires in place. This is a good way to keep the wires from getting tangled on things when you use The Hand.

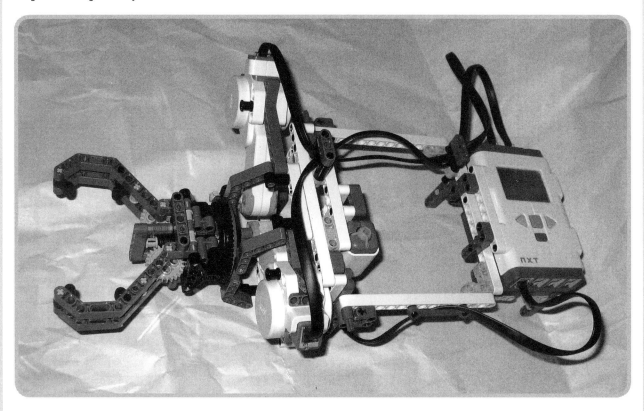

Figure 7-4: Cable connections for The Hand

programming the hand

Now we'll program The Hand to function. The program appears to be a little more complex than it really is. We basically have to make a few choices: Is the Touch Sensor pressed? Are the left or right arrows pressed?

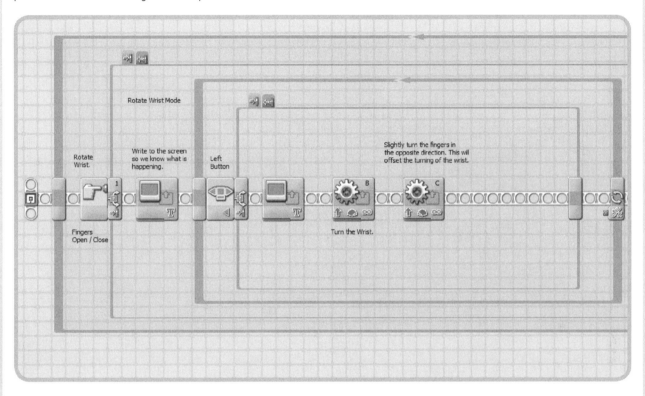

Figure 7-5: This portion of the program controls the following actions: The Touch Sensor is pressed (the wrist will turn); the left arrow is pressed (the wrist turns to the left).

Figure 7-6: Set the Loop block to run forever. This program never ends by itself. To end the program, simply press the dark gray button.

Figure 7-7: The Touch Sensor block controls which motor will be activated when you push the right and left arrow buttons. If the Touch Sensor is pressed, the wrist will rotate. If the Touch Sensor is not pressed, the fingers will open or close.

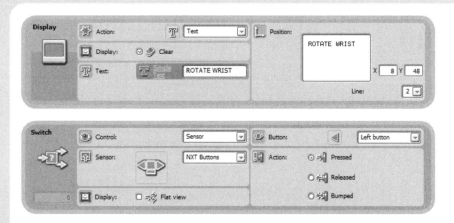

Figure 7-8: A Switch block—is the left arrow button pressed?

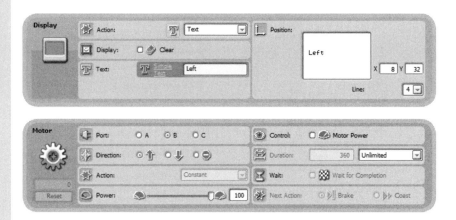

Figure 7-9: Run the motor at 100 percent power. This controls the wrist rotation.

Figure 7-10: Slightly rotate the center axle in the opposite direction. This prevents the fingers from opening or closing while the wrist is turning.

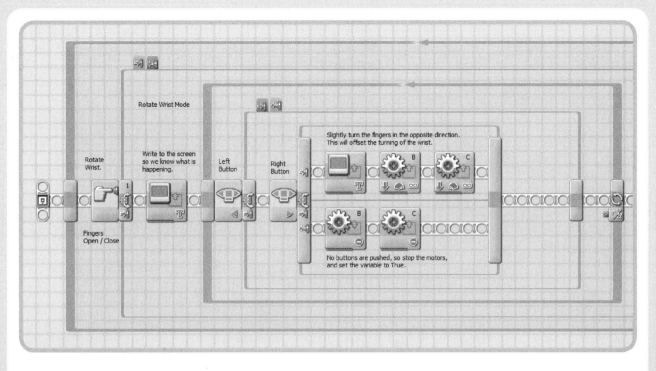

Figure 7-11: This portion of the program controls the following actions: The Touch Sensor is pressed (the wrist will turn); the right arrow is pressed (the wrist turns to the right).

Figure 7-12: A Switch block—is the right arrow button pressed?

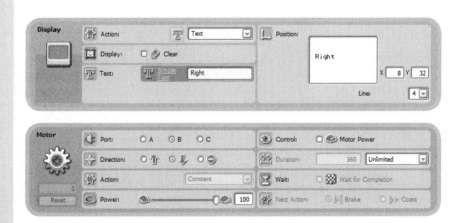

Figure 7-13: Run the motor at 100 percent power. This controls the wrist rotation.

Figure 7-14: Slightly rotate the center axle in the opposite direction. This prevents the fingers from opening or closing while the wrist is turning.

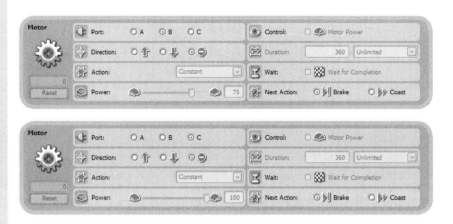

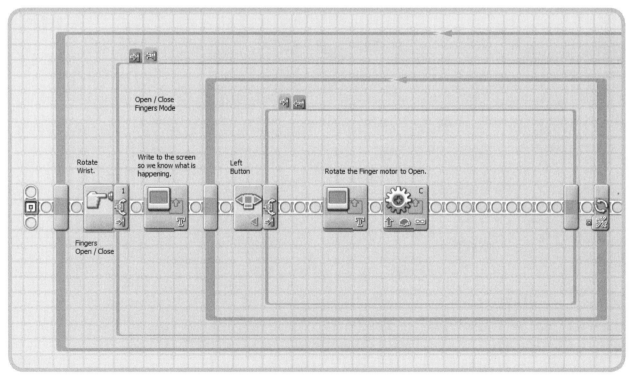

Figure 7-15: This portion of the program controls the following actions: The Touch Sensor is not pressed (you can control the fingers); the Left arrow is
pressed (the fingers will close).

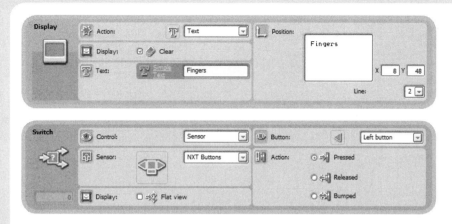

Figure 7-16: A Switch block—is the left arrow button pressed?

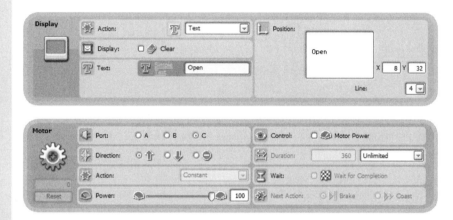

Figure 7-17: Run the motor at 100 percent power. This opens the fingers.

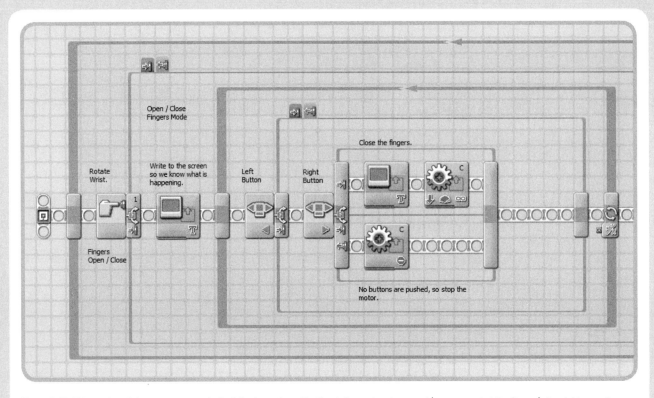

Figure 7-18: This portion of the program controls the following actions: The Touch Sensor is not pressed (you can control the fingers); the right arrow is pressed (the fingers will open).

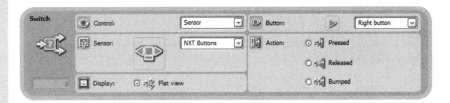

Figure 7-19: A Switch block—is the right arrow button pressed?

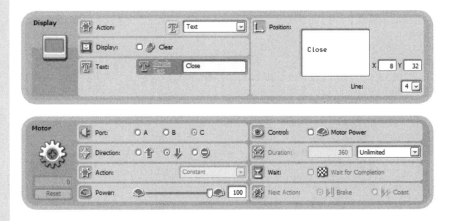

Figure 7-20: Run the motor at 100 percent power. This closes the fingers.

Figure 7-21: No buttons are pressed, so the motors stops.

further exploration

When you have built and programmed The Hand, experiment with it. A few suggestions for how you can modify both the structure and the programming are provided in the next sections.

building modifications for the hand

Here are some ways you can alter the construction of The Hand:

* Assemble The Hand so it can be worn on the left hand.
* Change the fingers so they can hold a pencil or other small object.
* If you open or close the fingers too far, the gear mechanism will pop loose. This is a result of using the worm gear (it is slow but very powerful). Try to design a finger structure that can withstand overtightening.
* Devise a way to determine when the fingers are fully open or closed.
* Rebuild the NXT Brick (that sits on your arm) so it will stay in place if you turn your hand upside down.

programming modifications for the hand

Try adding these capabilities to your program:

* Keep track of *full open* and *full close* so the program will stop opening or closing when that limit is reached.
* Rewrite the program using My Blocks. My Blocks will make the program easier to understand and are an excellent programming skill you can use in other programs.

8

SPC: the self-parking car

These days, we can't imagine life without cars; in fact, we use them every day and never give them a second thought. Just the same, learning to drive requires a lot of instruction and practice, which might make you wish for a car that parks itself. Some luxury cars today actually have special sensors that enable them to do this.

The Self-Parking Car is such a vehicle. Using an Ultrasonic Sensor, it measures the distances between the car and other objects and also measures to make sure the parking space is wide enough.

In this chapter, you'll learn how to build a robotic car that drives and steers like a real front-wheel drive car. Then we'll show you how to program it to park itself in response to your voice (or a clap of your hands).

Figure 8-1: The Self-Parking Car

building SPC

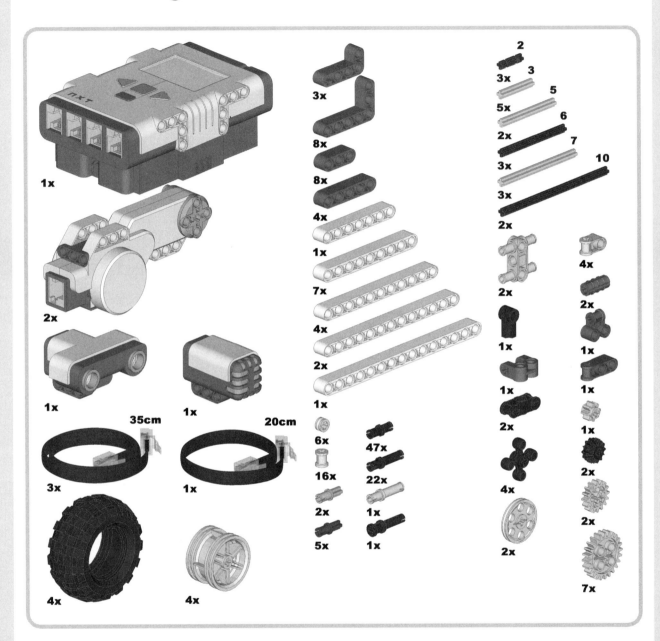

Figure 8-2: Bill of Materials

the front

1

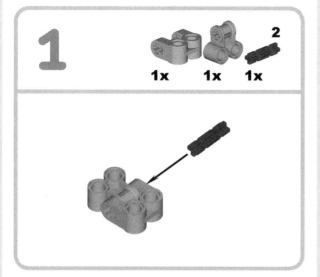

2

1x   1x   1x

2

2x

1x

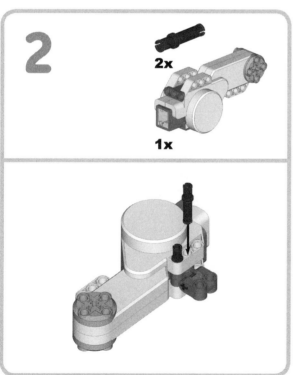

3

2x

1x

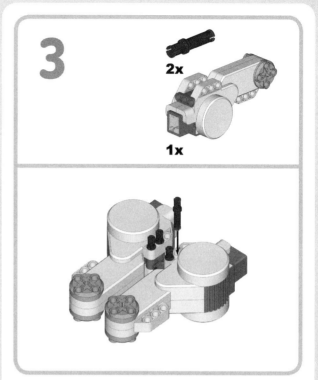

2x

1x

4

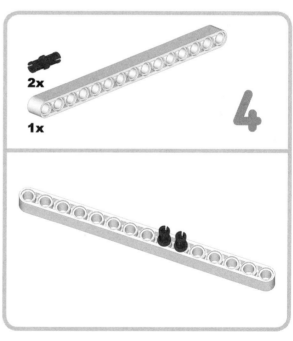

5

2x

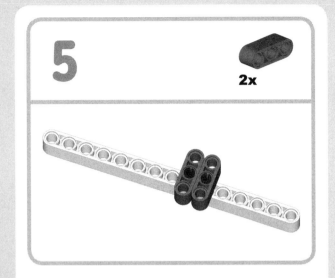

6

1x

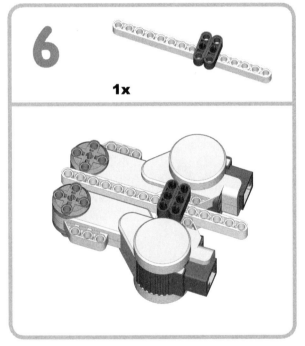

7

4x

2x

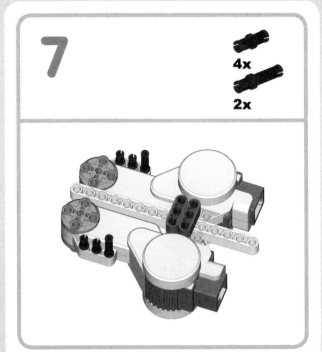

8

1x

2x

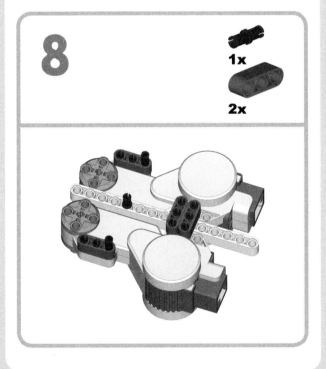

9

2x

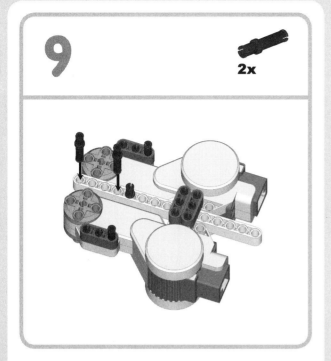

10

1x

1x

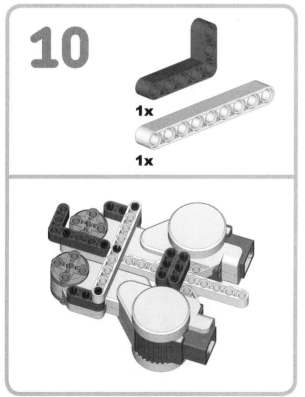

11

2x

1x

12

1x

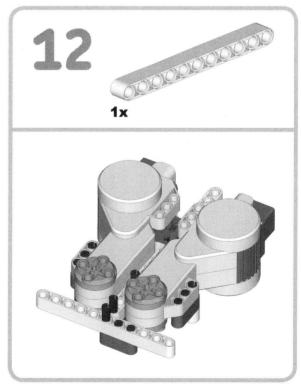

13

1x

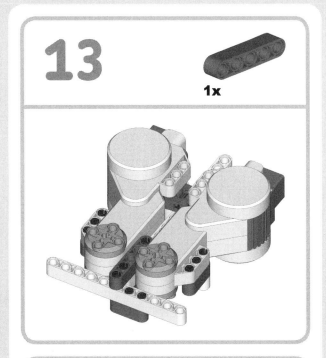

14

10

2x

2x

2x

4x

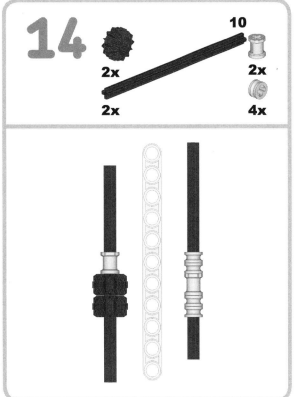

Use an 11-hole beam to measure the placement of the gears and bushings on the axles. You will not attach the beam to the car.

15

1x   1x

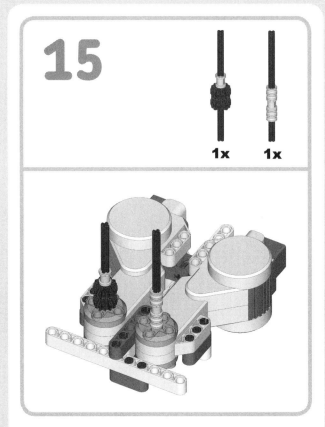

16

2x

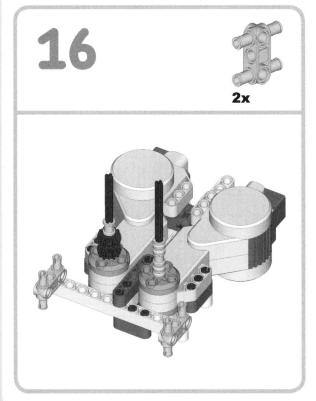

17

2x

5

1x

6

1x

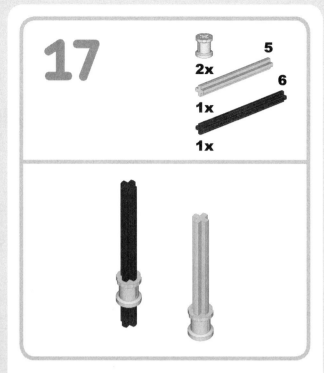

18

1x

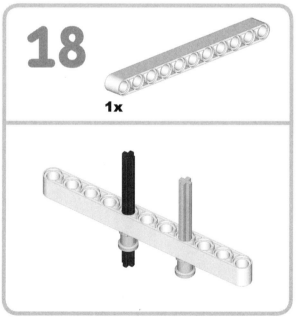

19

1x

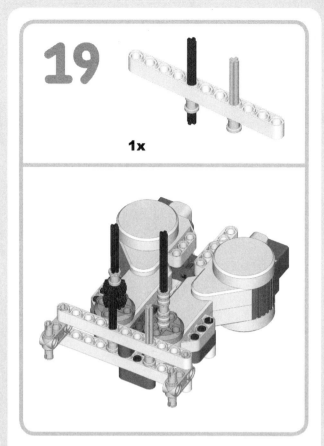

20

2x

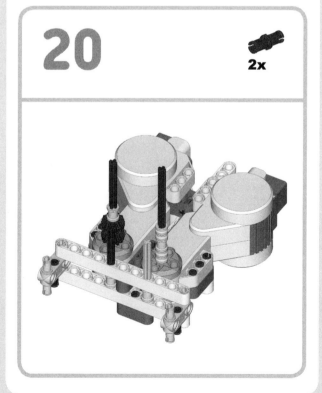

21

4x

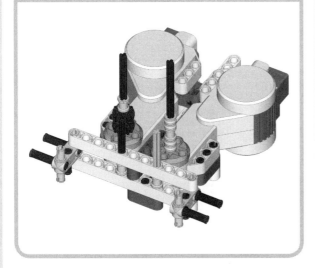

22

2x

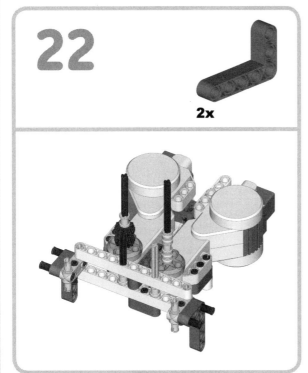

23

1x
1x
2x

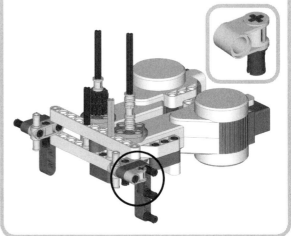

24

1x
1x
2x

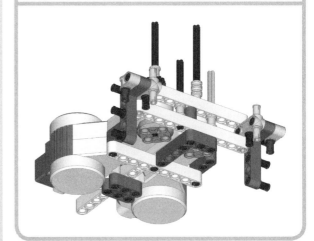

25

2x

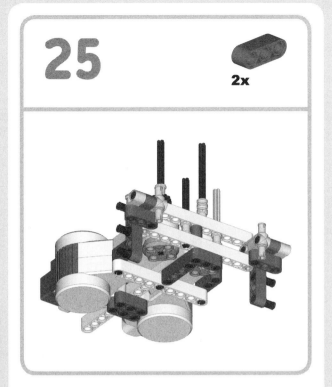

26

2x

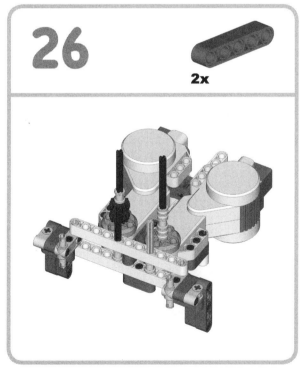

27

2x   7

1x   1x   1x

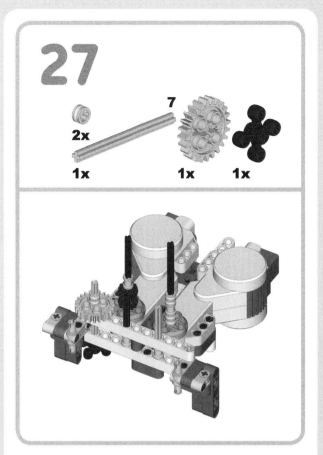

28

1x   5

1x   1x

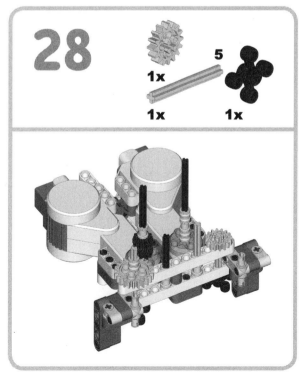

29

1x

2x

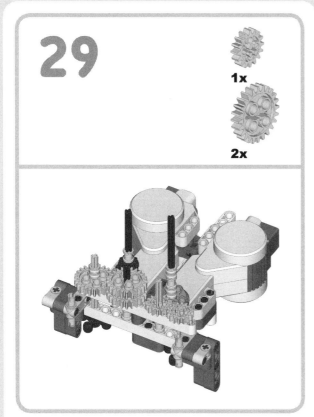

30

2x   1x

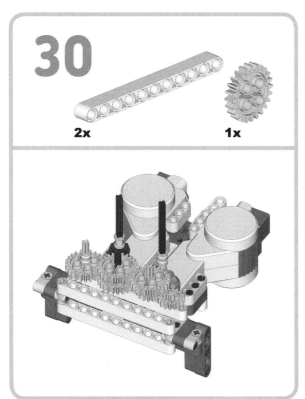

31

2x

1x

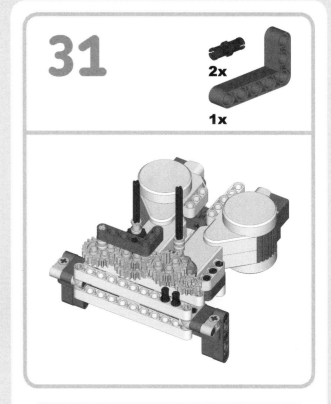

32

1x

1x

1x

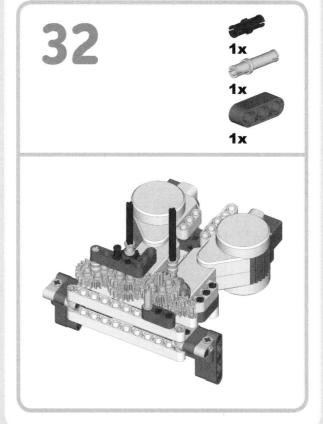

33

1x
1x

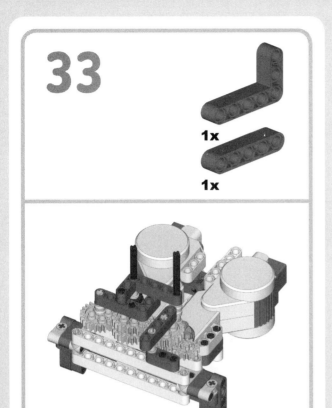

34

2x
1x

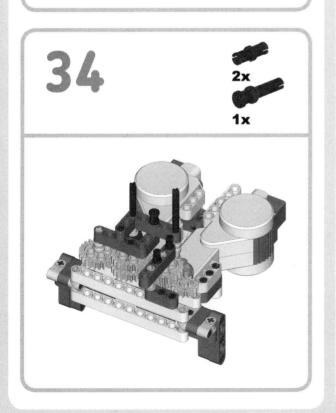

35

3
1x 1x 1x

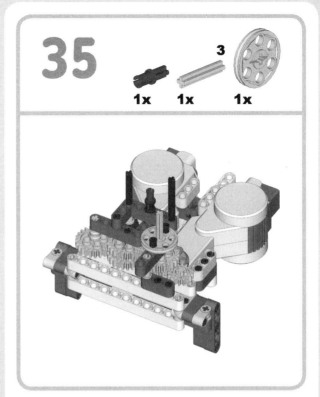

36

1x
1x 1x

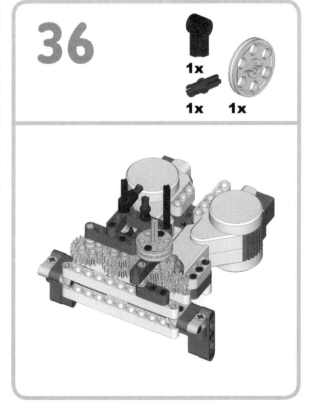

37

1x

1x

1x

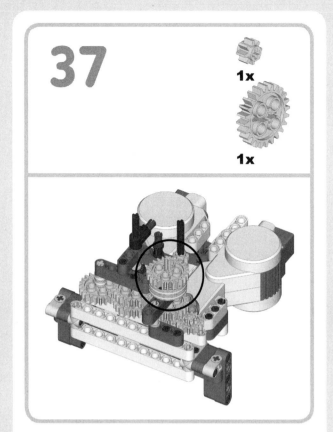

38

1x

1x

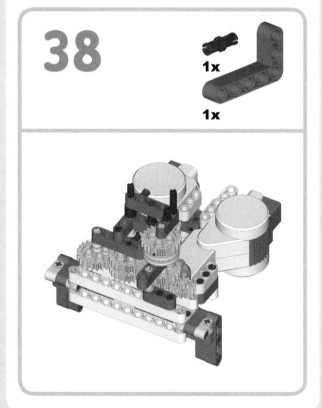

39

1x

1x

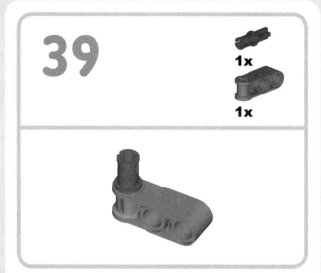

40

2x

1x

41

1x

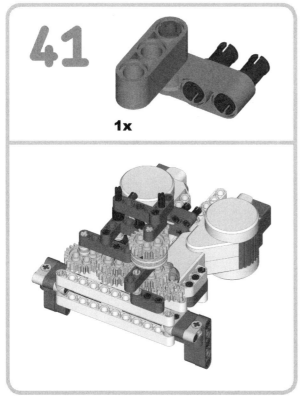

42

2x

43

6

2x

2x

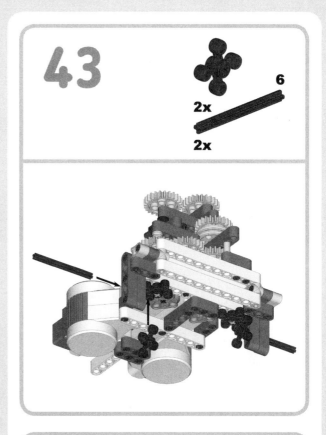

44

2x

45

2x

2x   2x

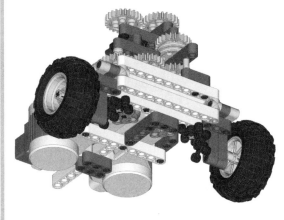

the brick and ultrasonic sensor

1

4x

1x

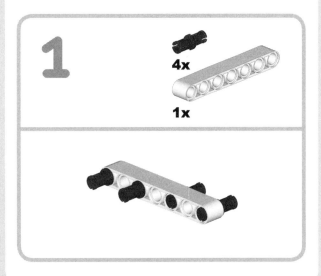

2

1x

1x

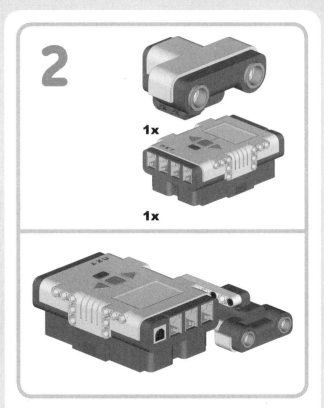

3

1x

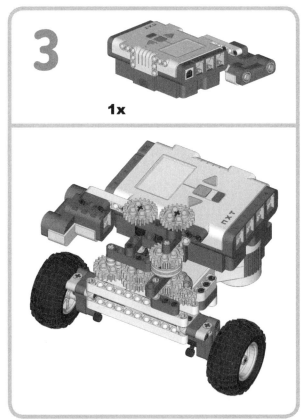

the rear

1

10x

2x

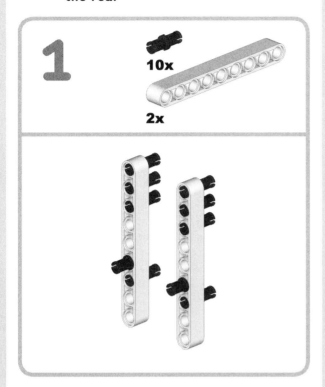

2

4x

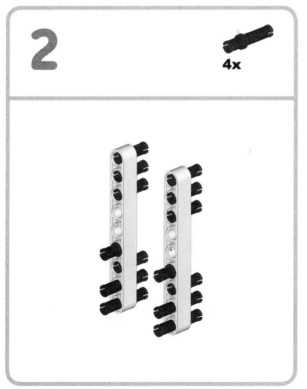

3

2x

1x

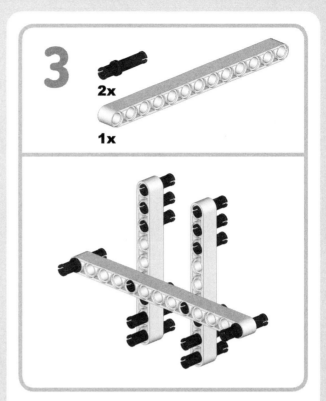

4

2x

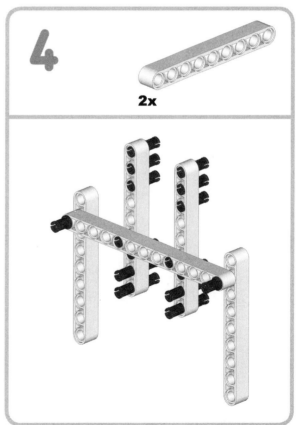

5

2x

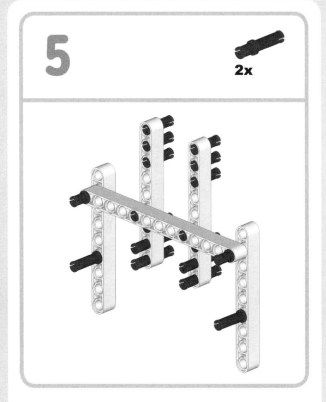

7

4x

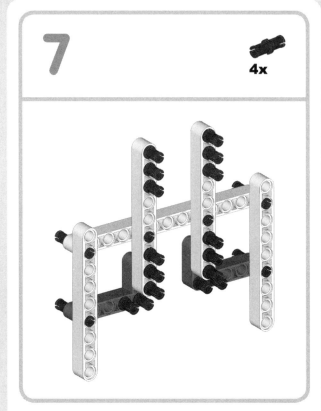

6

2x

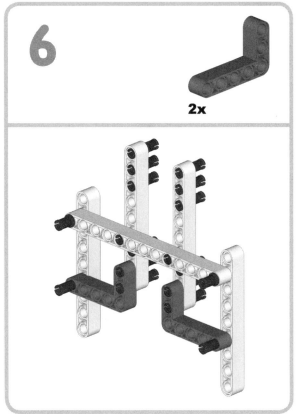

8

2x

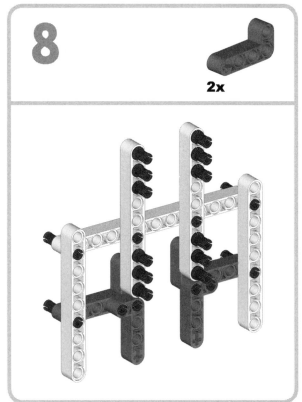

9

4x

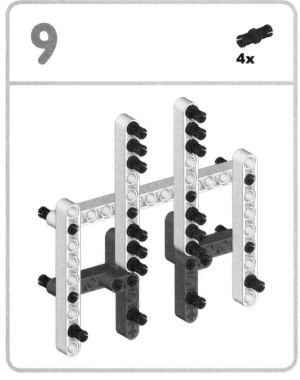

10

1x

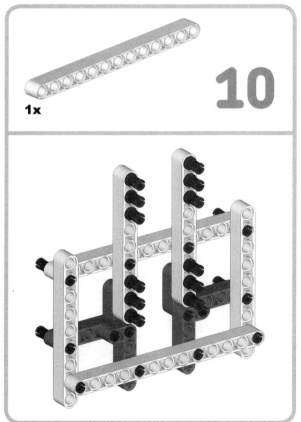

11

2
2x

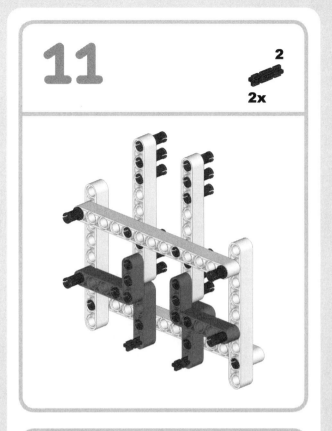

12

2x

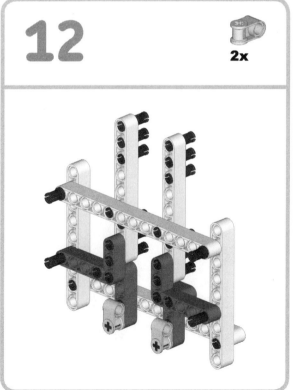

13

2x

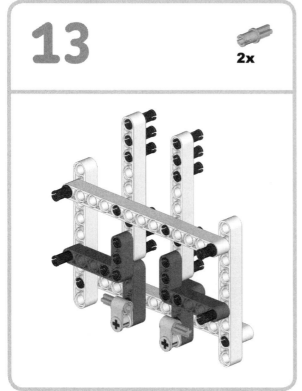

14

2x

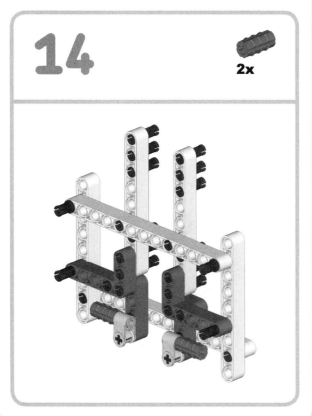

15

2x
7
2x

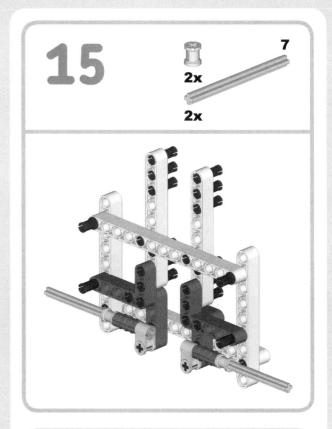

16

2x

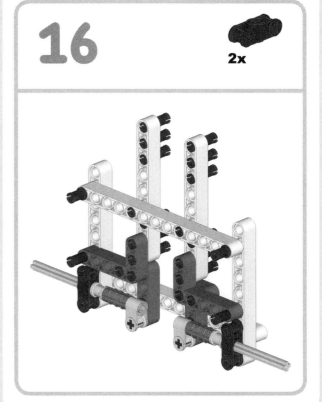

17

3
4x

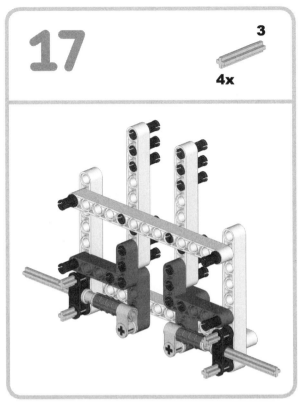

18

4x

2x

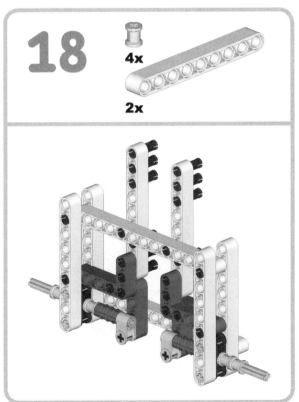

19

2x

2x

2x

2x

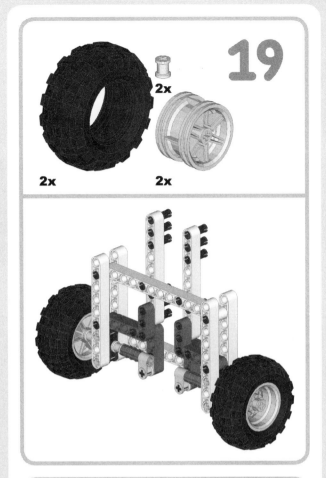

20

1x   1x

21

1x

1x

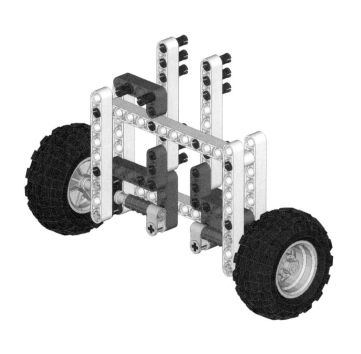

22

1x

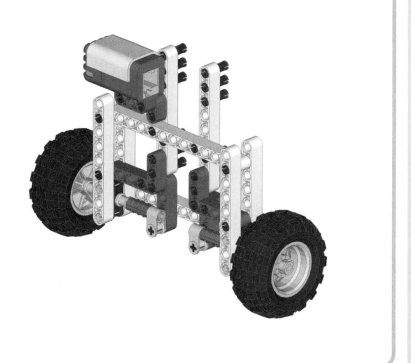

23

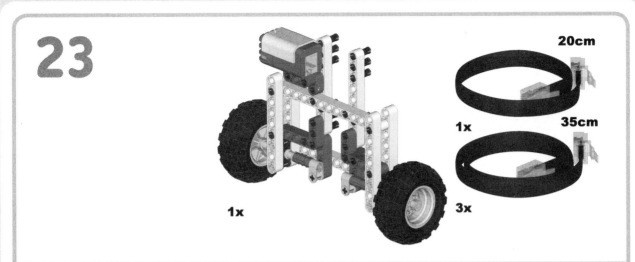

20cm

1x

35cm

1x

3x

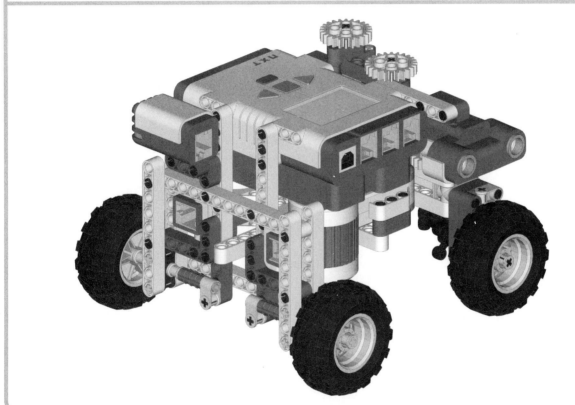

connecting the cables

Congratulations! You've finished building the Self-Parking Car. When you've connected the cables, you'll be ready to program it.

If you look at the last building step, you'll see a motor on the left that enables the car to steer, which we'll call the Steering Motor; we will refer to the one on the right as the Driving Motor.

Using medium-length (35 cm/14 inch) cables, connect the Steering Motor to output port B on the NXT Brick, the Driving Motor to output port A, and the Sound Sensor to input port 2. Use a short cable (20 cm/8 inch) to connect the Ultrasonic Sensor to input port 4.

programming SPC

With this program, the vehicle will park itself between two solid objects. Before starting the program, place the car parallel to the first object, as shown in Figure 8-3.

When you make a loud sound, triggering the program to start, the car begins to move while constantly looking for a space big enough to place itself in. When it finds one, it will guide itself into the parking space with a sequence of preprogrammed actions. If there's not enough room, it will stop and apologize by saying "Sorry!"

To make the program readable and simple, we'll use three My Blocks that will control steering. One will make the car steer to the left, another to the right, and the third one will make it go straight ahead.

my block #1—steer right

To make the vehicle steer right, the Steering Motor must turn 70 degrees in a reverse direction. These 70 degrees must be measured from the point where the steering wheel is in the middle so that we can steer to the right point, independent of the current position of the steering wheel.

To do this, we use a Switch block to determine if the steering wheel is already in the correct position. If it is, the Switch block will activate the *true* statement, which stops the steering motor. If this is not the case (*false*), it activates the motor. This switch is continuously repeated and therefore, in the end, the Steering Motor will stop when it has reached the correct position, and the car now steers right.

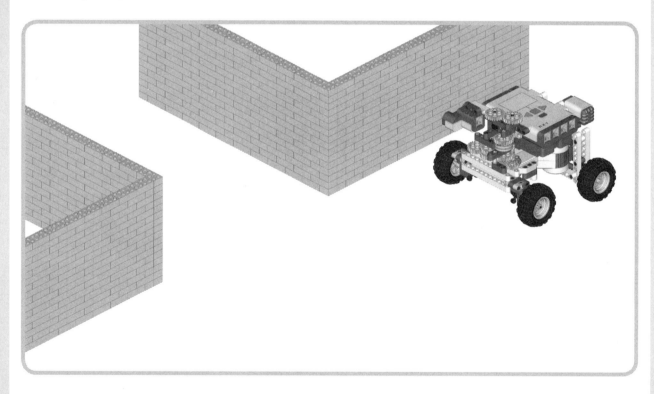

Figure 8-3: Place the car parallel to the first object as shown here.

Step 1: Create a new document and place five blocks configured as shown in Figure 8-4.

Select all the program blocks that we've just added and create a My Block called Steer Right. (Select an appropriate icon too, if you like.) Delete the My Block from your work space so you can create the next one.

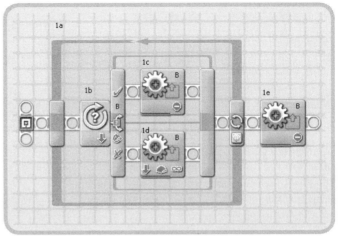

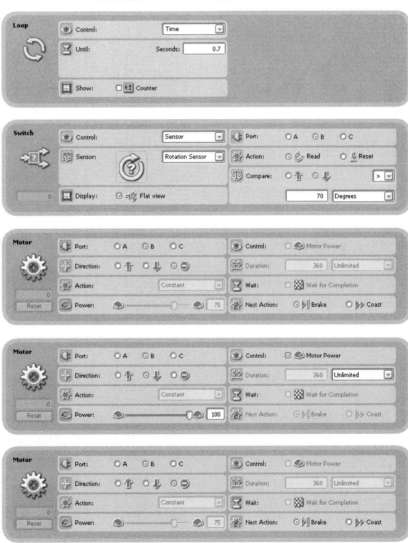

Figure 8-4: The configuration of the blocks for the Steer Right My Block

my block #2—steer left

The Steer Left My Block works just like the Steer Right My Block, except that the Steering Motor moves in the opposite direction and the Switch block is configured to measure rotation in the appropriate direction.

Step 2: Place five blocks configured as shown in Figure 8-5.

Now select all the program blocks that we added and turn them into a new My Block called Steer Left. (Select an appropriate icon too, if you like.) Delete the My Block from your work space so you can create the next one.

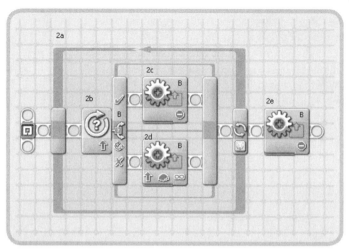

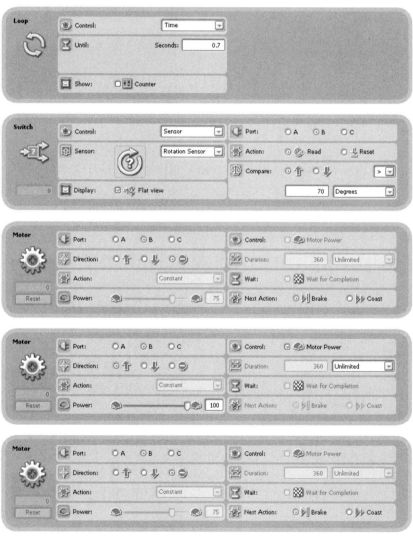

Figure 8-5: The configuration of the blocks for the Steer Left My Block

my block #3—steer forward

The Steer Forward My Block makes the Steering Motor rotate back to the middle, making the car drive straight ahead. This block is different from the other two My Blocks you just created because now there are two possible directions for the Steering Motor to turn—left or right.

When the car is already moving straight ahead and the Steer Forward My Block is used, we don't want the Steering Motor to move. We can check this by looking at the Rotation Sensor. If the Rotation Sensor gives us a value close to zero, the Steering Motor should not move.

After this check, we'll use Switch blocks to determine the position of the steering wheels. Then we'll make the Steering Motor turn in the appropriate direction until the Steering Motor returns to its original position—the middle.

Step 3: Place three blocks configured as shown in Figure 8-6. The value of the Rotation Sensor is wired into the Move block.

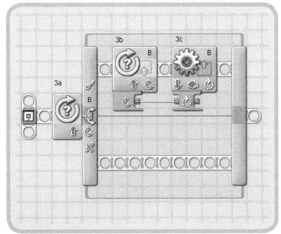

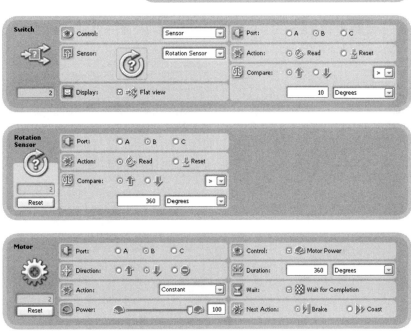

Figure 8-6: The configuration of the blocks placed in step 3

Step 4: Place and wire four blocks configured as shown in Figure 8-7. The value of the Rotation Sensor is wired into the Move block.

Now select all the program blocks that we added and turn them into a new My Block called Steer Forward. You must then delete the My Block from your work space so you can create the final program.

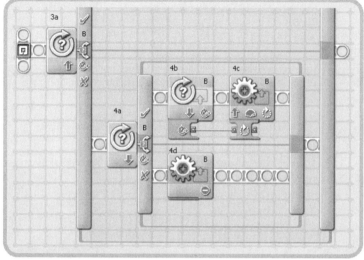

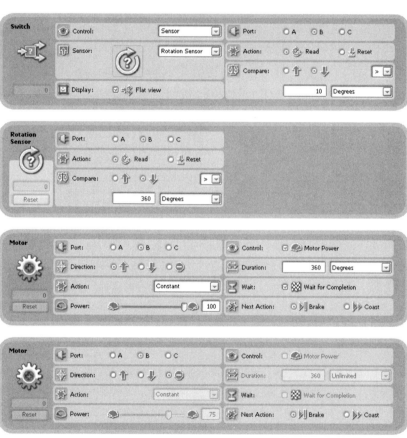

Figure 8-7: The configuration of the blocks placed in step 4

296   CHAPTER 8

main program

When you start the program, the car will wait for a loud sound to tell it to start moving. When the program begins, the car keeps moving until the Ultrasonic Sensor measures a distance greater than 6 inches (15 cm), which tells the robot there is space to park.

It then resets the Rotation Sensor in the Driving Motor and moves on until it sees something (the end of the parking space). Then it compares the value of the Rotation Sensor (the distance traveled) to 1,200 degrees, the minimum distance between the two objects. If this is greater than 1,200, there is enough space for the car to park itself. This will be demonstrated in steps 5 and 6.

Step 5: Place four blocks configured as shown in Figure 8-8.

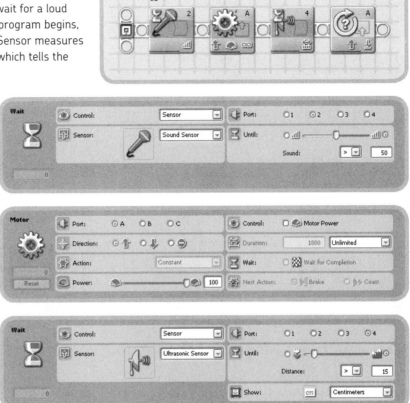

Figure 8-8: The configuration of the blocks placed in step 5

Step 6: Place five blocks configured as shown in Figure 8-9. The value of the Rotation Sensor is wired into the Compare block. The true/false output is wired into the Switch block.

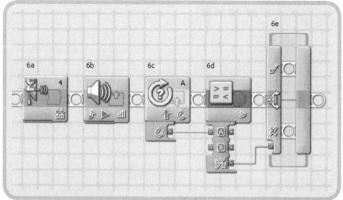

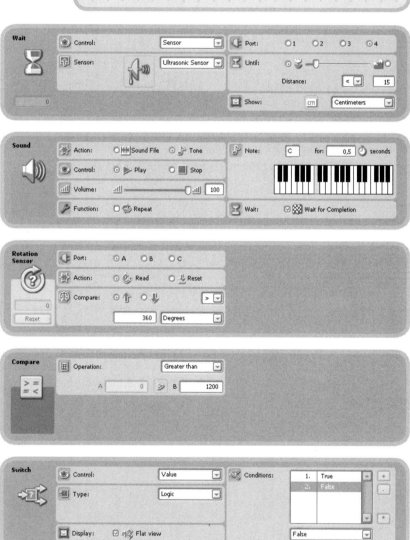

Figure 8-9: The configuration of the blocks placed in step 6

Steps 7, 8, and 9 will accomplish certain tasks, depending on whether the Compare block finds a distance that is
long enough for the robot to park between the objects.

If the result is *false*, the space is not big enough, and
step 7 will make the car stop and say "Sorry!" If the result is
true, there is enough space to park, so the blocks placed in
steps 8 and 9 will guide the car into the gap with a sequence
of Move blocks and Steering My Blocks.

Step 7: Place two blocks configured as shown in
Figure 8-10.

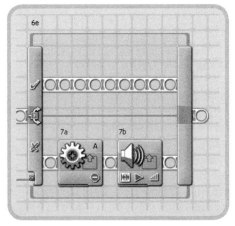

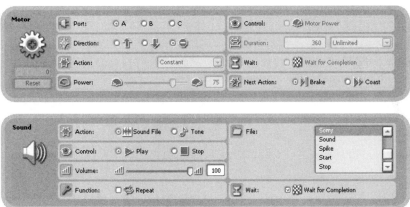

Figure 8-10: The configuration of the blocks placed in step 7

Step 8: Place five blocks configured as shown in Figure 8-11.

NOTE After the second object
is detected, the car will not stop
moving immediately. Before parking, it has to move a bit more so
that it doesn't crash into something. This is accomplished by
Motor block 8a. If your robot
doesn't work well, gets stuck, or
hits one of the objects, you can
try modifying the value in the
box labeled *duration*, which in my
program is set to 0.7 rotations. A
value greater than 0.7 will make
the car park slightly closer to the
second object. A value less than 0.7
will do the opposite, moving the
car closer to the first object.

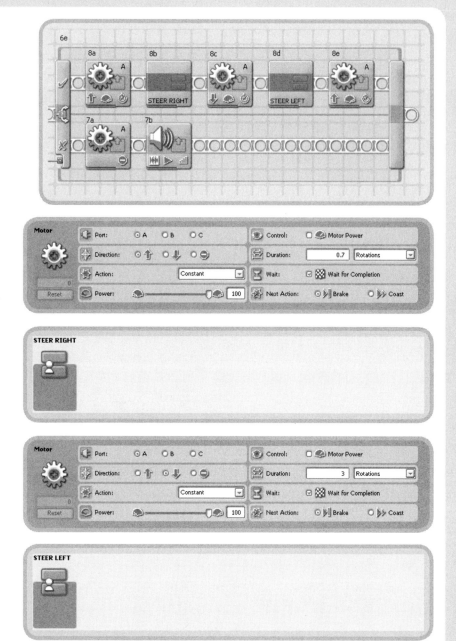

Figure 8-11: The configuration of the blocks placed in step 8

Step 9: Place five blocks configured as shown in Figure 8-12.

You can now download the program to your NXT Brick. Remember to save your program.

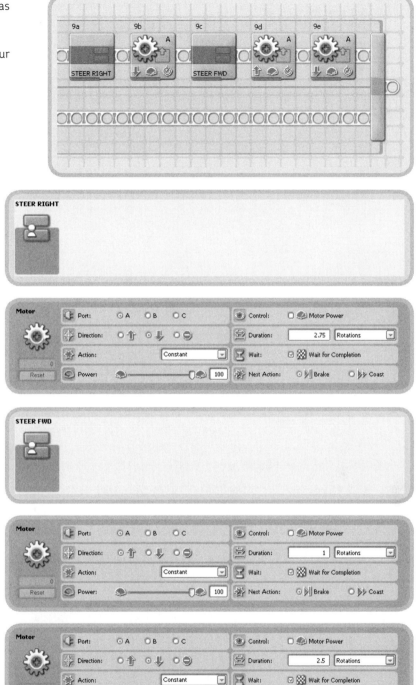

Figure 8-12: The configuration of the blocks placed in step 9

further exploration

In this chapter you learned to create the My Blocks that control the steering of the car. By combining these My Blocks with other blocks, you can easily program your vehicle to avoid walls. This can be done by programming it to steer and back up when you clap your hands or by mounting the Ultrasonic Sensor in the front of the car to sense walls and other objects.

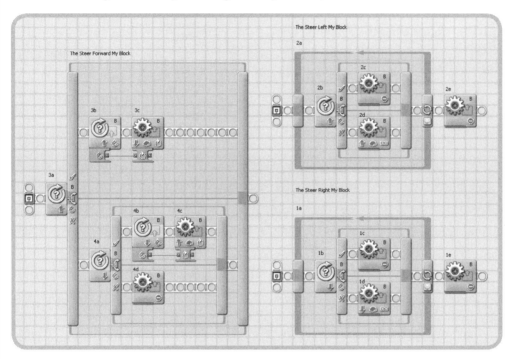

Figure 8-13: This is an overview of the three My Blocks you've created to control the steering motor.

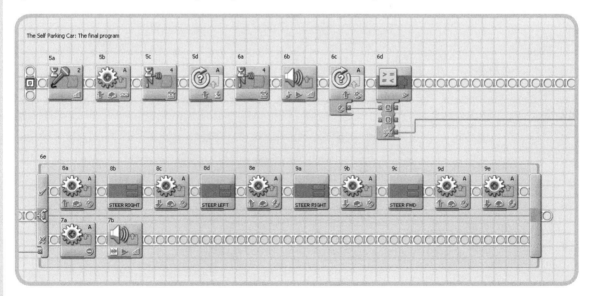

Figure 8-14: This is the final program that enables the SPC to park between two objects. (The program has been split in two here for readability.)

GrabBot: a robot that finds, grabs, lifts, and moves

This chapter will teach you how to build and program a robot that can find and grab objects by itself. Using the Ultrasonic Sensor, GrabBot spins around looking for the closest object, moves toward it, rescans for a more precise location, then grabs and lifts the object.

The unique physical aspect of this robot is the grabber, which controls two functions with a single motor: grabbing and lifting. A Touch Sensor is used to detect if the grabber is lifted all the way up.

While you can pick up many things with the grabber, I have had the greatest success using a plastic cup with a diameter of about 2.5 inches (7 cm) and a height of about 4 inches (10 cm).

Whatever you use, keep in mind that the wider an object is, the easier it will be for the robot to find it. Square objects, like the pile of bricks in Figure 9-2, can only be found if one of the flat sides of the structure faces the robot.

Figure 9-1: GrabBot

Figure 9-2: Three objects (left to right): a pile of standard LEGO bricks, a structure made from parts left in the NXT kit, and a plastic cup

building GrabBot

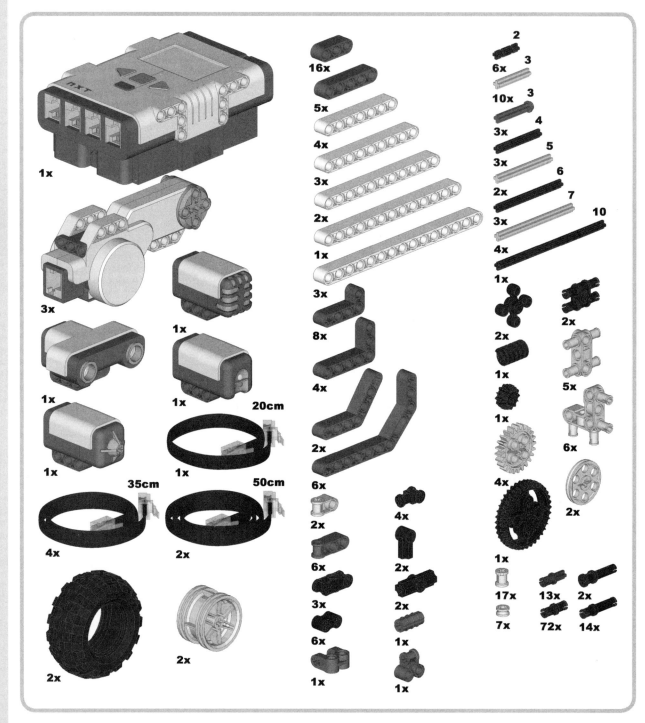

Figure 9-3: Bill of Materials

the grabber

1

2x
4
2x

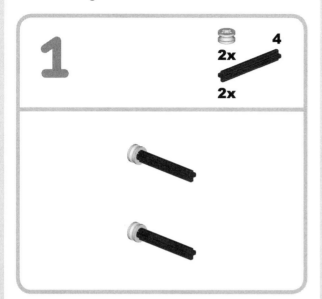

2

1x
1x

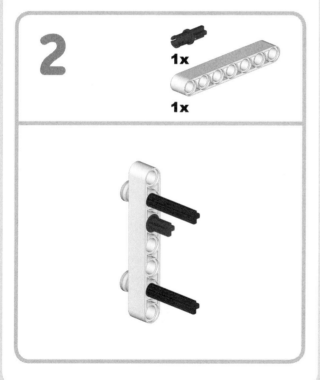

3

1x
2x
1x

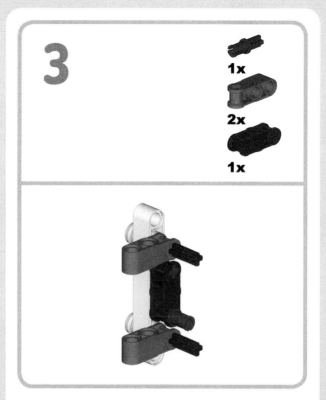

4

2x
1x

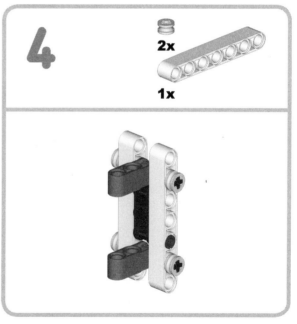

5

4x

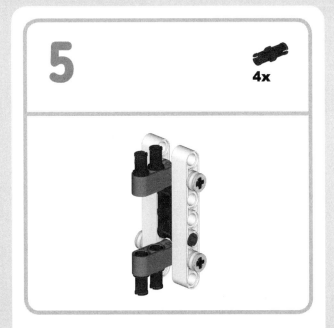

6

1x

7

1x   1x

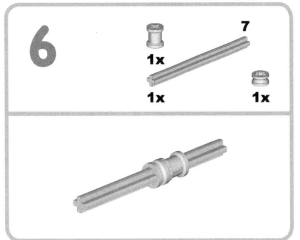

7

1x   1x

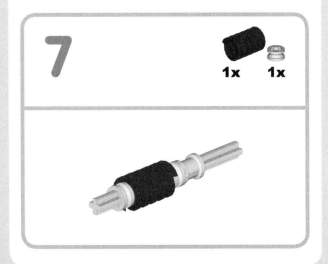

8

1x   1x

1x

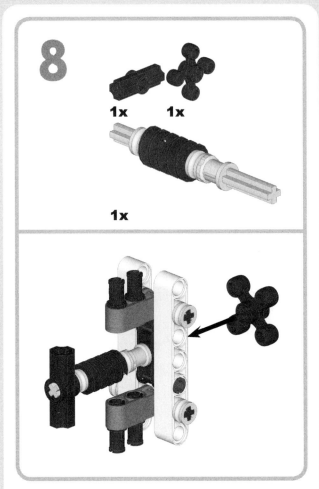

9

1x

1x

10

1x **1x** **3** **1x**

Build two of these.

11

1x

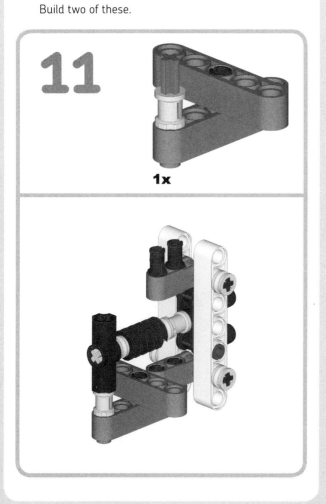

12

1x **1x** **7** **1x**

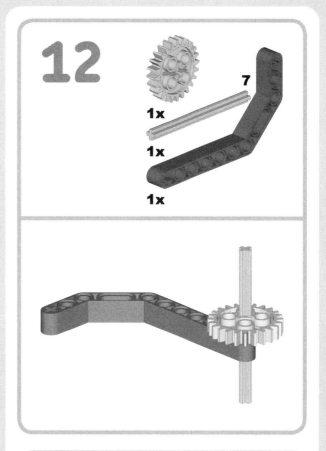

13

1x **3** **1x** **1x**

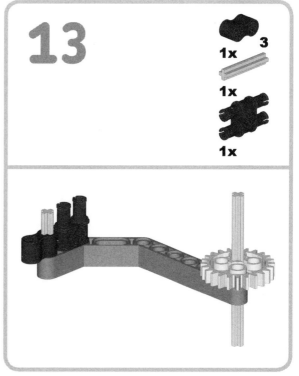

14

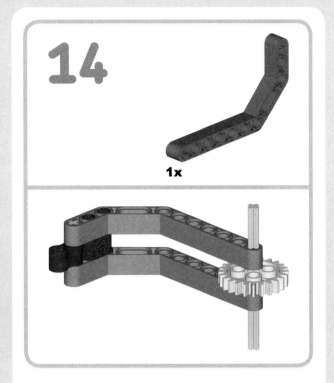

1x

Build two of these.

15

2x

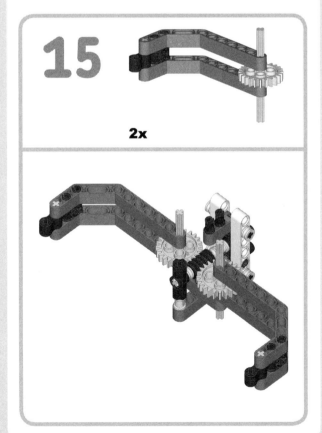

16

1x   4x

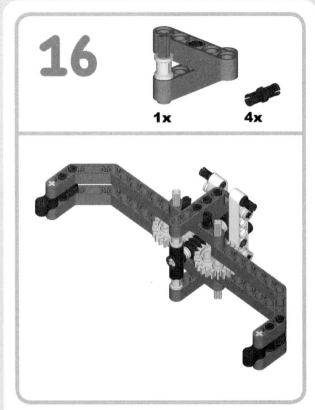

17

2x

2x

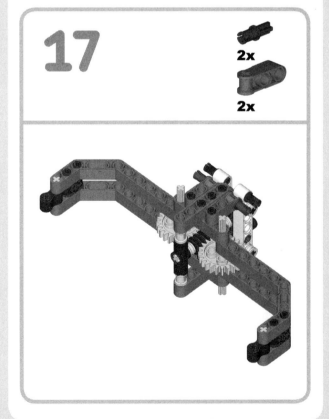

18

2x 4x

19

4
1x
5
1x
1x

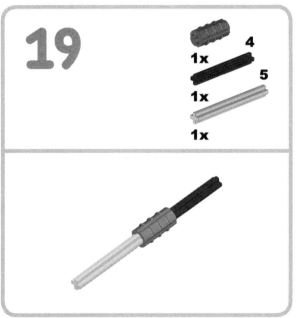

20

1x

21

2x
1x

22

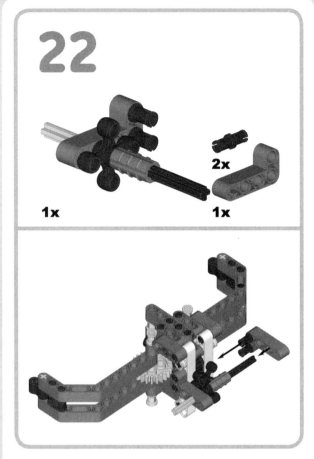

2x

1x   **1x**

23

1x   **3**

1x

24

2x   **2**

1x

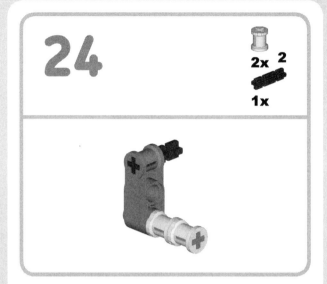

25

1x   **1x**

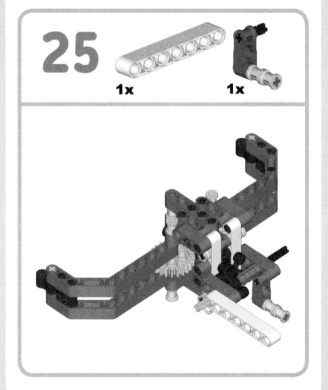

26

1x

the grabber motor

1

3x

1x 10

1x 2x

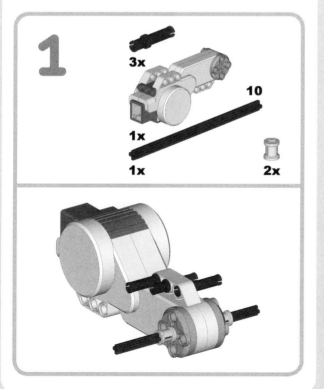

2

2x

2x

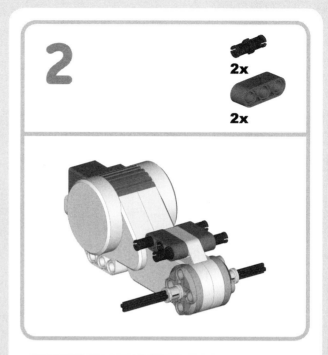

3

1x

2x 2x

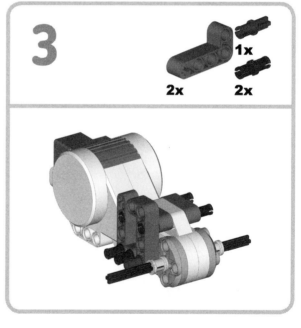

4

1x **1x**

1x

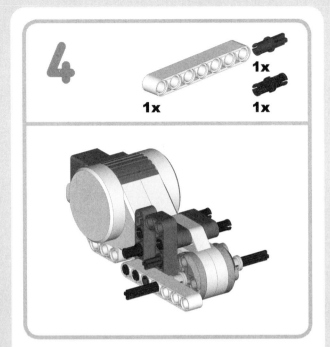

6

2x **2x**

2x **3x**

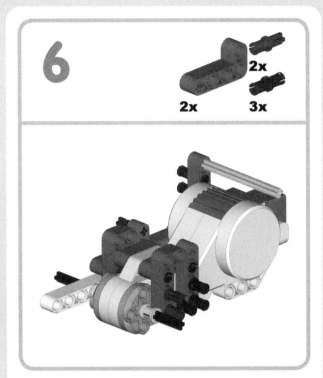

5

2

1x

2x

1x **7**

1x

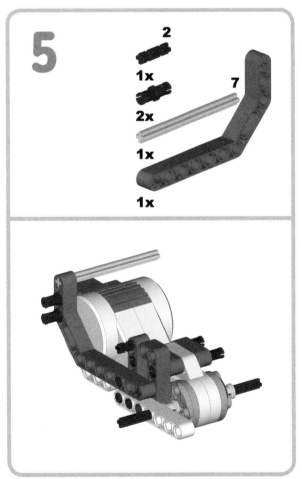

7

2

1x

2x

1x

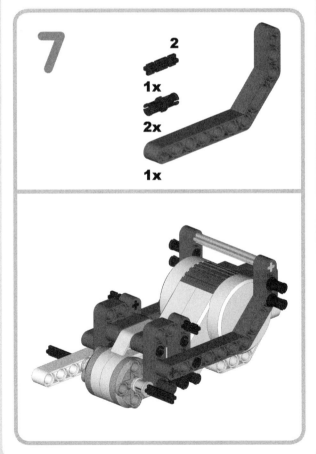

8

6x 2x

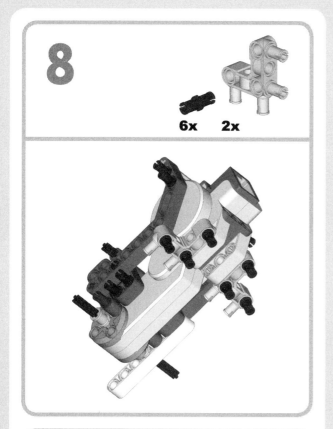

9

2x

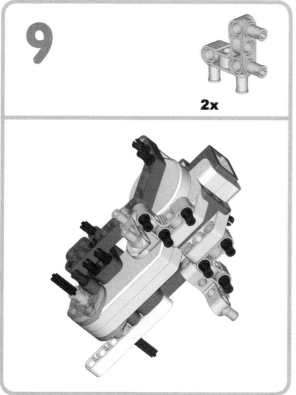

10

1x

1x

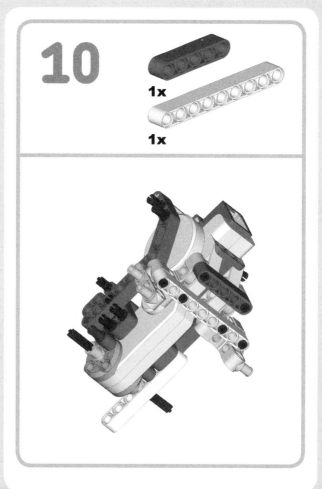

11

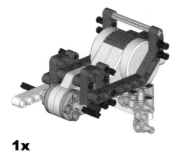

1x

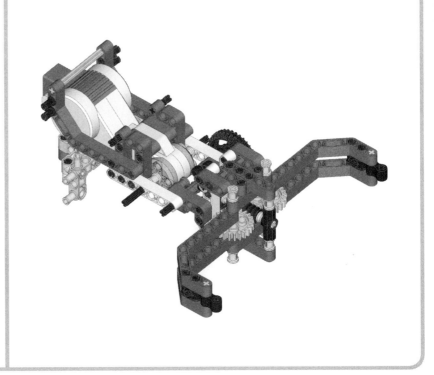

12

2x   1x

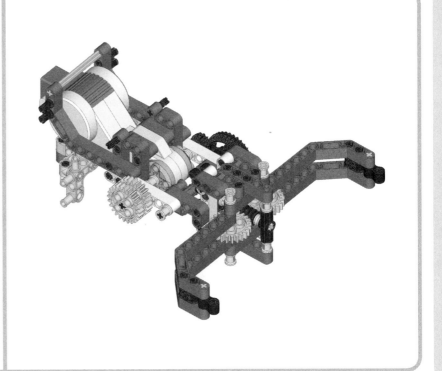

13

2x

1x

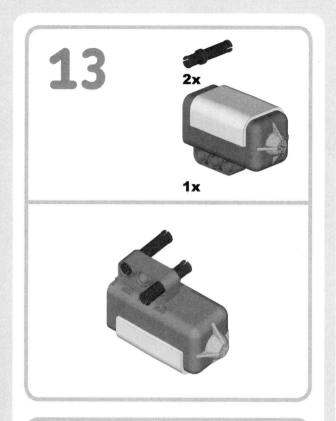

14

2x

15

1x   2x

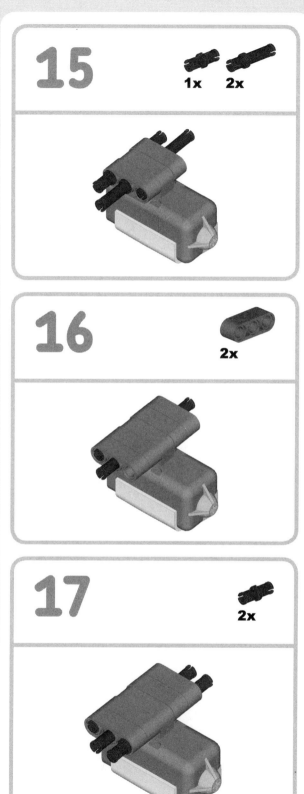

16

2x

17

2x

18

1x

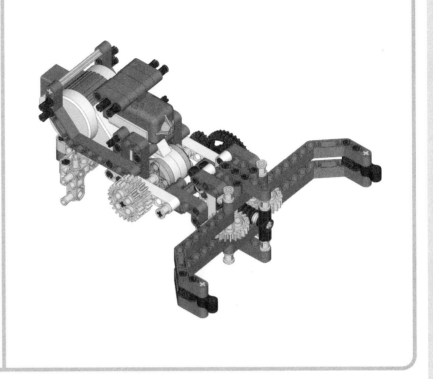

19

4x

2x

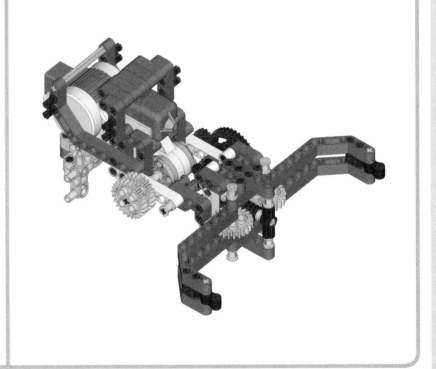

the driving motors

1

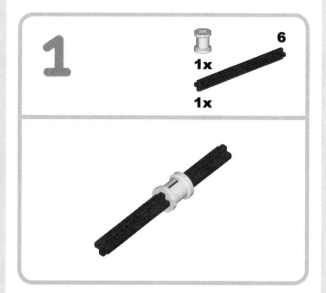

2

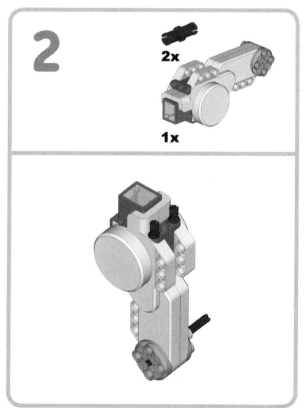

3

4

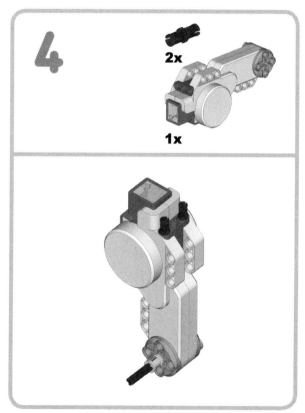

5

1x   1x

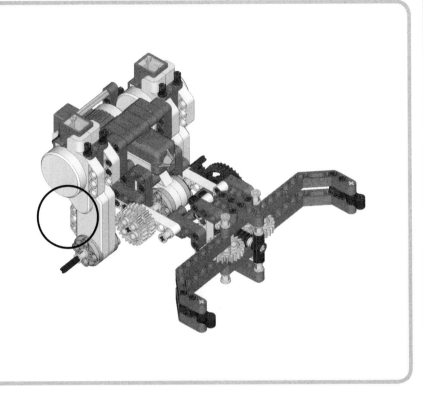

6

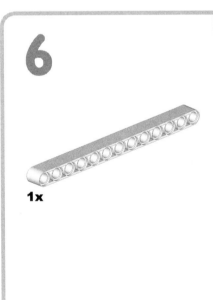

1x

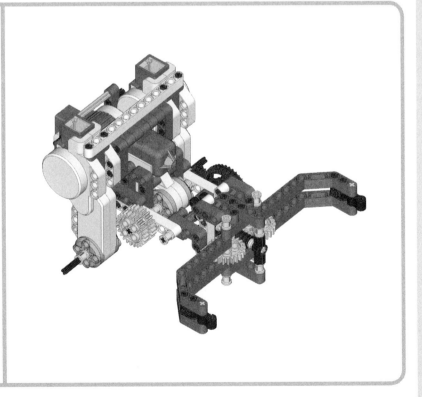

the rear frame

1

2x   1x

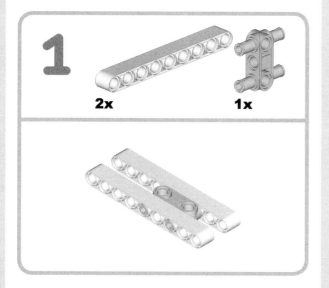

2

4x

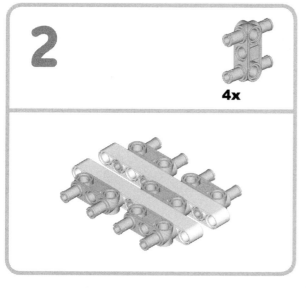

3

2x

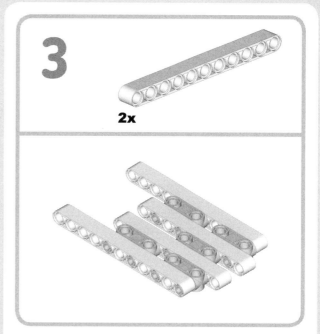

4

6x

1x

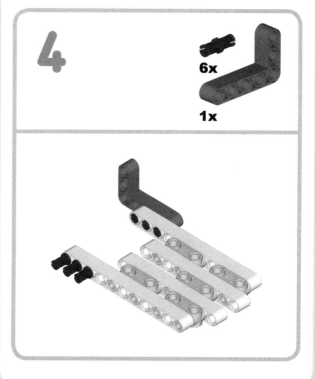

5

1x

1x

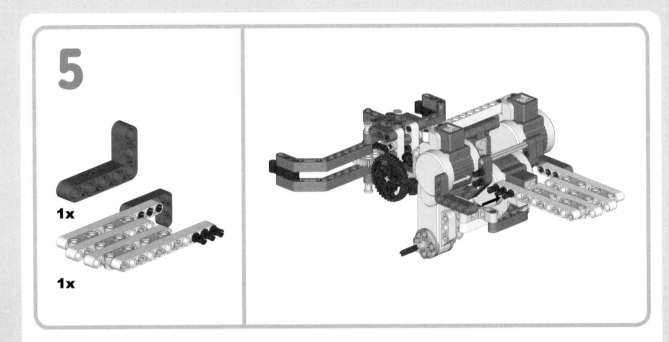

the caster wheel

1

1x   5   3

1x   1x

1x   1x

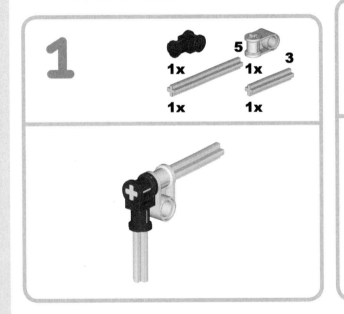

2

1x   1x   3

1x   1x

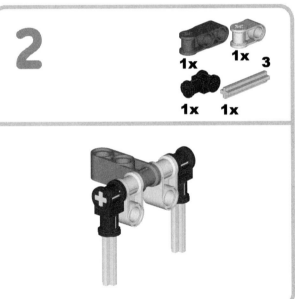

3

2x
2x

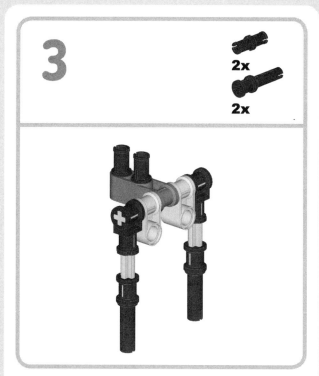

4

2x
2x

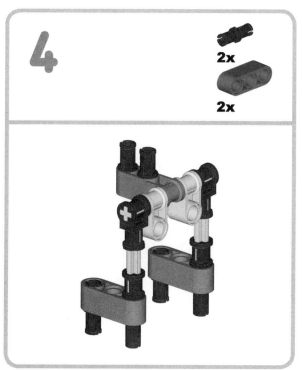

5

4x
1x

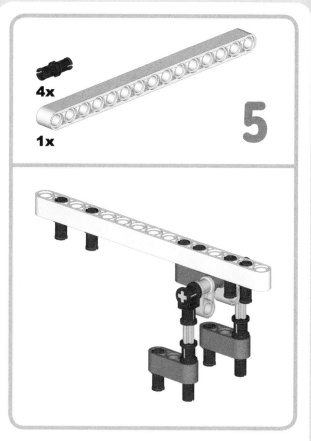

6

1x
1x
6
1x
1x

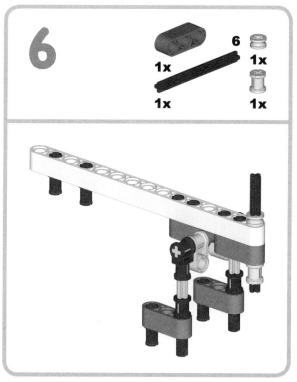

7

2

2x   1x   2x

8

1x

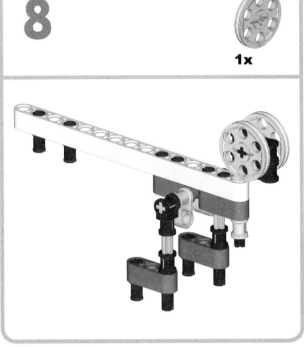

9

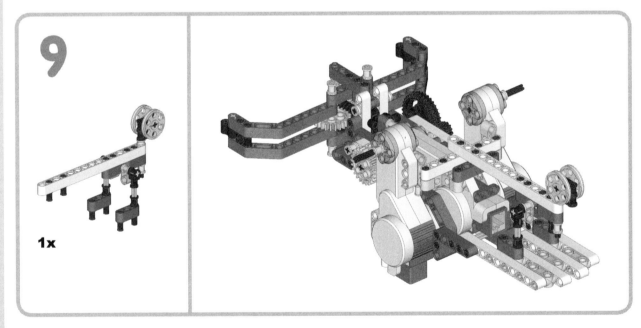

1x

the ultrasonic sensor

1

2x 1x 1x

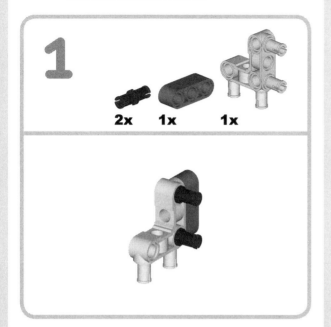

2

2x

1x

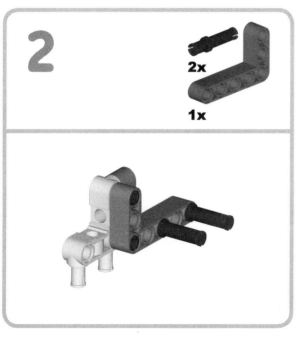

3

1x

35cm

1x

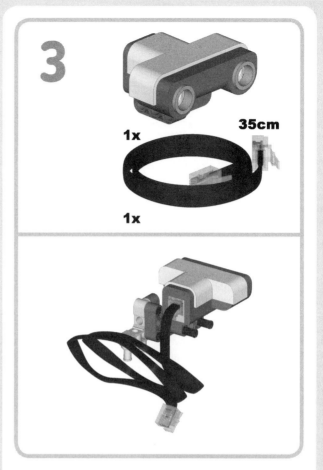

4

2x

1x

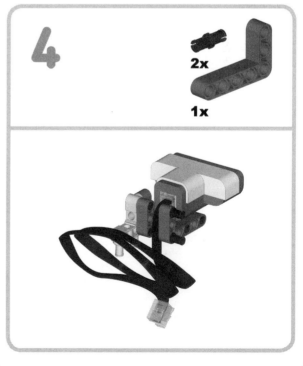

5

1x

1x

Attach the Ultrasonic Sensor subassembly and wrap the cable around the frame.

The short (20 cm/8 inch) cable connects to the motor in the middle, while the medium-length (35 cm/14 inch) cable connects to the Touch Sensor. You'll need to wrap the medium cable around the frame.

6

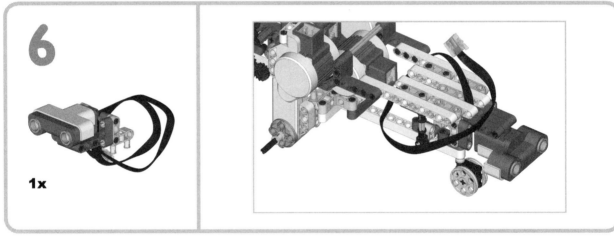

1x

7

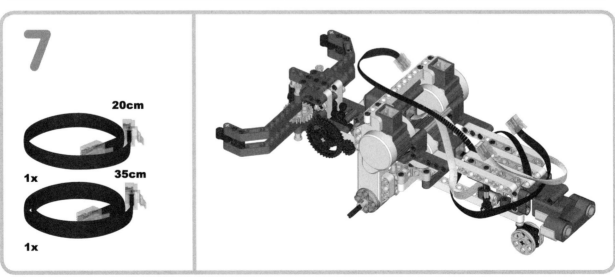

20cm

1x

35cm

1x

the brick

1

2x
1x
1x

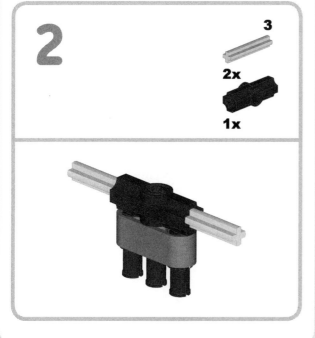

2

3
2x
1x

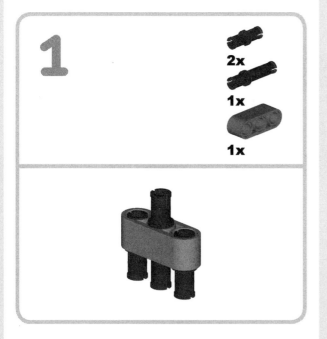

3

2x
2x

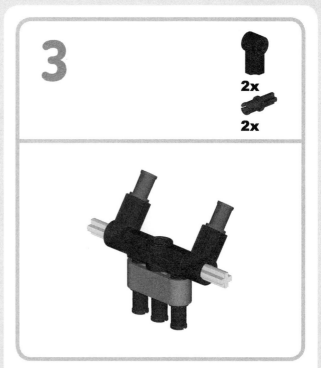

4

2x
2x

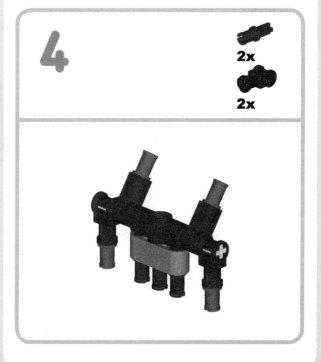

5

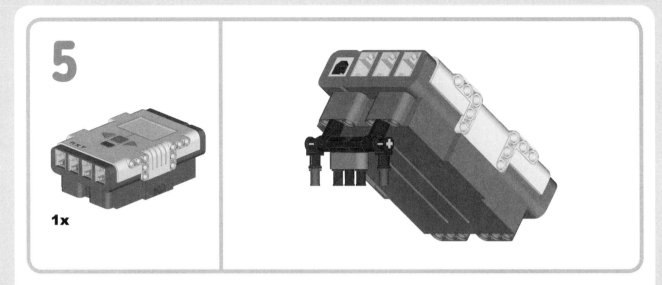

1x

Attach the Brick subassembly as shown. Then connect the Touch Sensor to input port 1, the Ultrasonic Sensor to input port 4, and the middle motor to output port A.

6

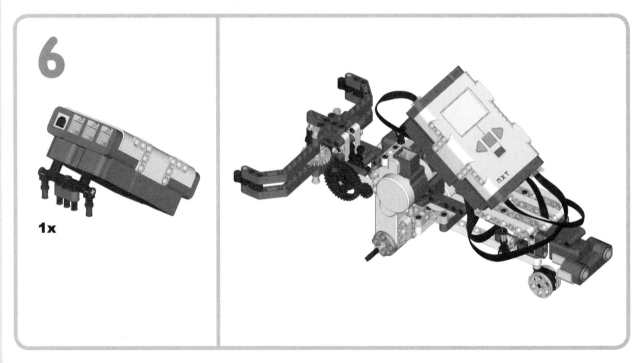

1x

the sound sensor

1

3
1x
1x

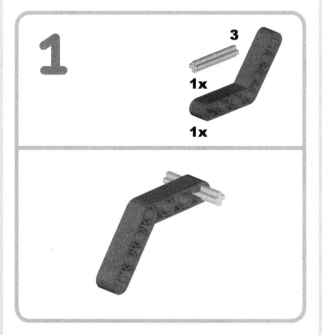

2

2x 3
3x

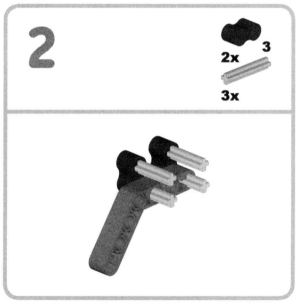

3

1x

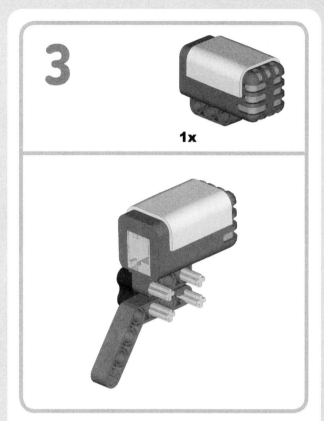

4

1x 2x 1x

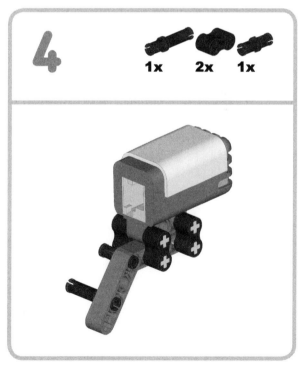

5

1x
1x

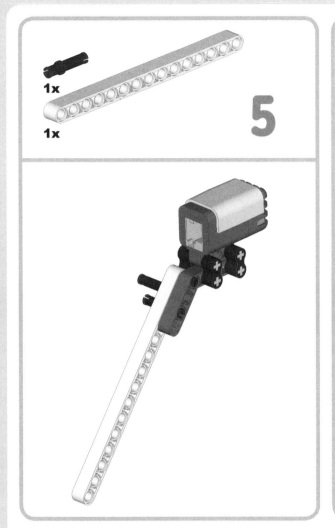

6

2x
1x

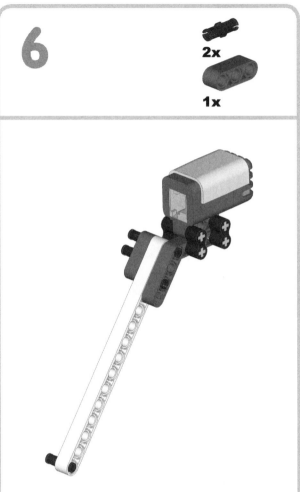

7

1x

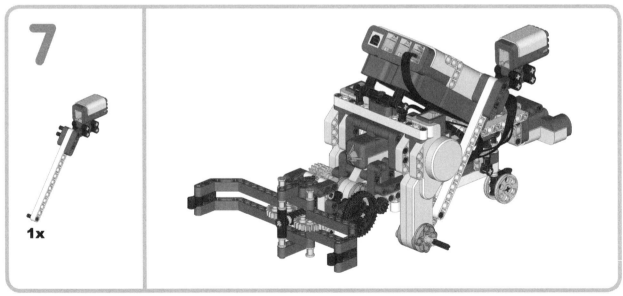

The NXT kit includes sixteen of these TECHNIC bushings, but some kits have eighteen of them (two are reserve
parts). If you have only sixteen of these available, you could
leave out the bushings that would normally be placed on the
outer side of the wheel axles.

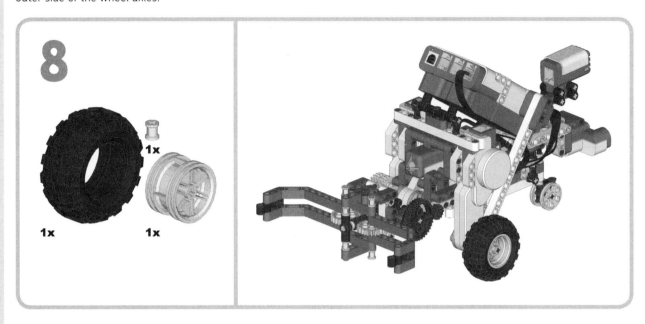

the light sensor

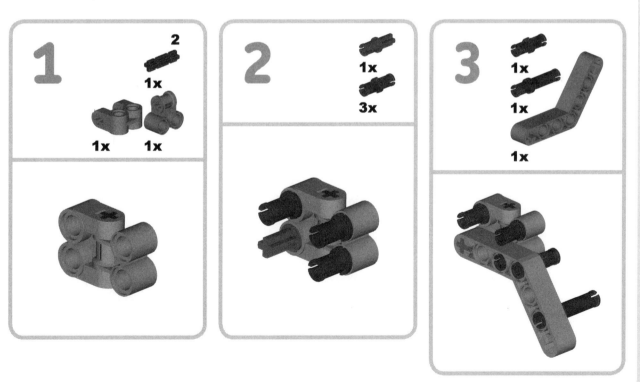

4

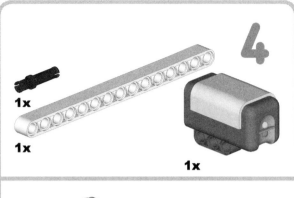

1x
1x
1x

5

2x
1x

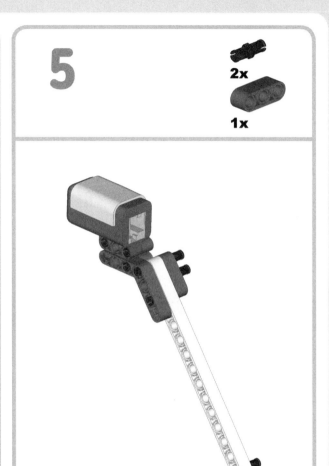

6

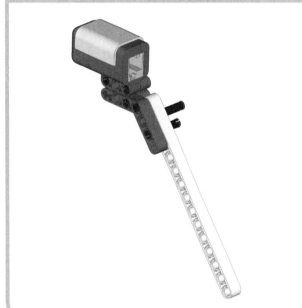

1x

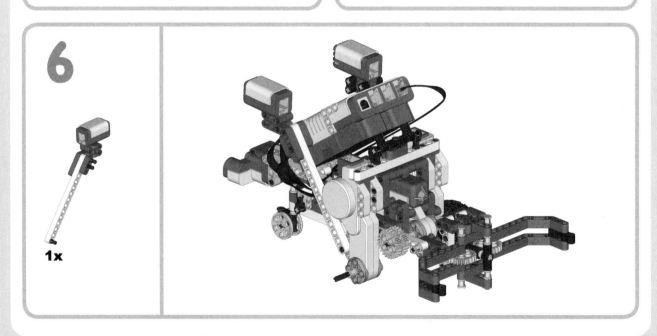

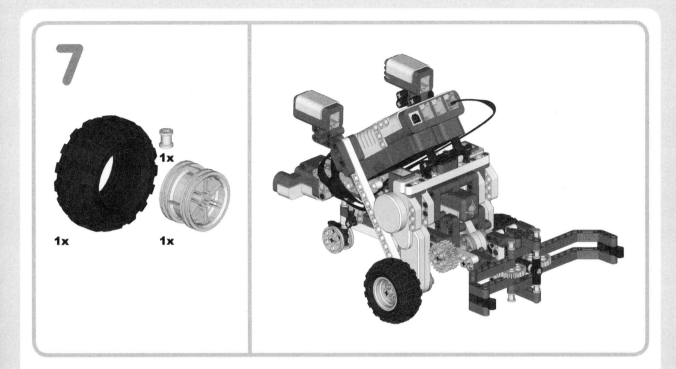

7

1x 1x 1x

connecting the cables

Using two medium-length (35 cm/14 inch) cables, connect the Sound Sensor to input port 2 and the Light Sensor to input port 3. Use two long (50 cm/20 inch) cables to connect the left motor to output port B and the right motor to output port C.

35cm

2x

50cm

2x

programming GrabBot

In this program, we want to make GrabBot find, grab, lift, and move objects around the room while avoiding walls. When we clap our hands or make another loud sound, the robot should drop its cargo and stop moving.

my block—search

To make the program easier to understand, we'll create a My Block, which we'll use twice in our final GrabBot program. This miniprogram rotates the robot 180 degrees while the Ultrasonic Sensor looks for the closest object in range. The Rotation Sensor that is built in to the motor tracks the direction in which the object was seen. When the scanning is complete, the robot turns to face the direction of the object.

Figure 9-4 illustrates the scan. In this case the values 25 (the lowest measured sensor value) and 255 (the value of the Rotation Sensor at that point) are stored in GrabBot's memory. Note that in reality, the robot will make hundreds of measurements instead of just four.

Storing the lowest measured value along with the Rotation Sensor value works as shown in Figure 9-5.

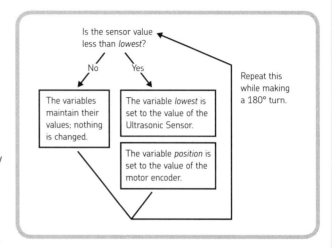

Figure 9-5: Program flow for the Search My Block

GrabBot repeats this procedure until it has turned 180 degrees. Then it rotates in the opposite direction and stops at the heading stored in the variable named *position*.

Now we'll create the Search My Block.

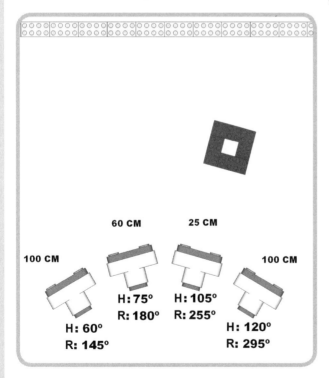

Figure 9-4: Four measurements of the scan: H stands for Heading (the angle the robot has turned compared to its starting point), and R stands for the value of the Rotation Sensor in the NXT motor.

Step 1: Create a new document. Declare two number variables, *lowest* and *position*, and place five blocks configured as shown in Figure 9-6.

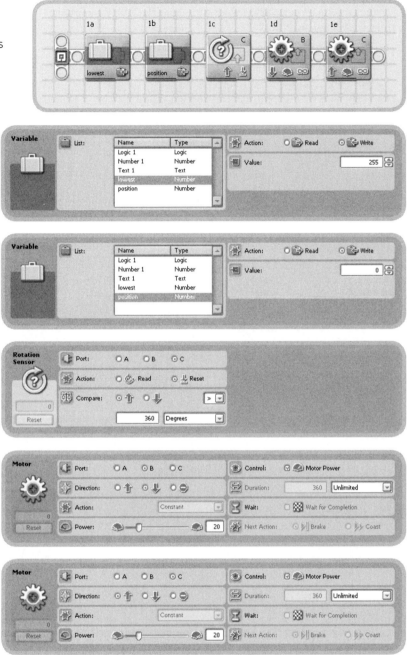

Figure 9-6: This is the configuration of the blocks placed in step 1.

Step 2: Place a Loop block and a Switch block configured as shown in Figure 9-7.

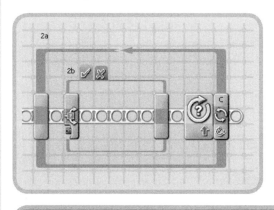

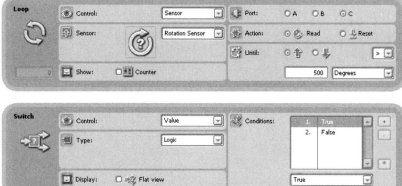

Figure 9-7: This is the configuration of the blocks placed in step 2.

Step 3: Place three blocks configured as shown in Figure 9-8.

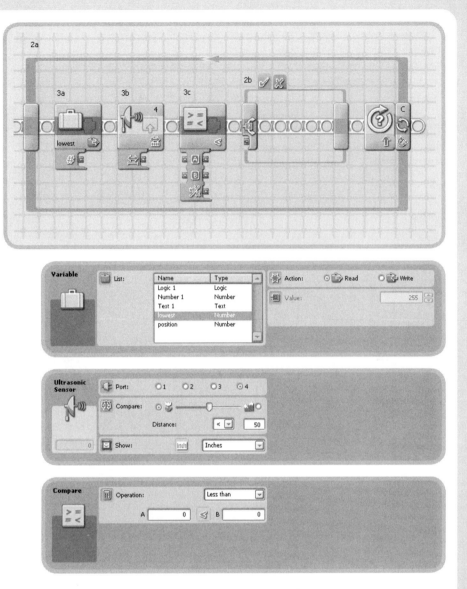

Figure 9-8: This is the configuration of the blocks placed in step 3.

Step 4: Place three blocks configured as shown in Figure 9-9. The tab under *X* (false) is empty.

Step 5: Wire the blocks as shown in Figure 9-10.

NOTE The wire between blocks 3b and 3c and the wire between 3c and 4a are two separate wires, not one single, long wire.

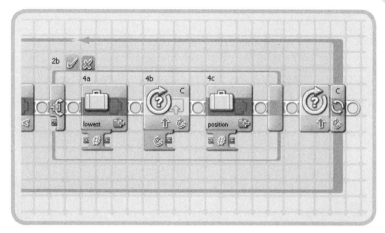

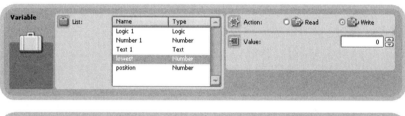

Figure 9-9: This is the configuration of the blocks placed in step 4.

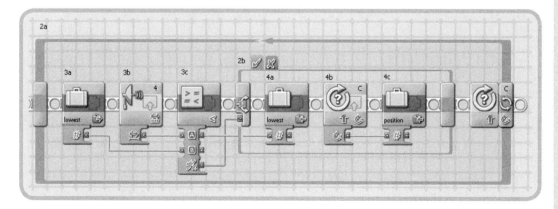

Figure 9-10: This is the wire configuration in step 5.

Step 6: Place three blocks configured as shown in Figure 9-11.

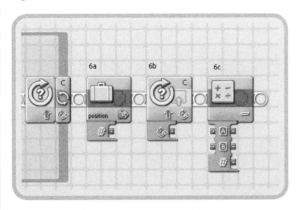

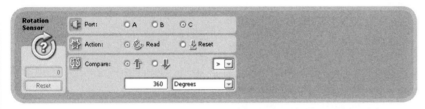

Figure 9-11: This is the configuration of the blocks placed in step 6.

Step 7: Place three blocks configured as shown in Figure 9-12.

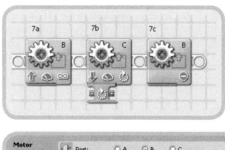

Figure 9-12: This is the configuration of the blocks placed in step 7.

Step 8: Wire the blocks as shown in Figure 9-13.

Now select all the program blocks that we added and turn them into a new My Block called *Search*. (Select an appropriate icon too, if you like.)

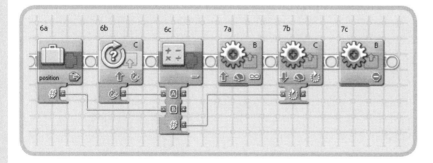

Figure 9-13: This is the wire configuration in step 8.

main program

To make the program accurate, we will scan twice. After the first scan, GrabBot will move forward a certain distance based on the value in the *lowest* variable. The further the object is away, the more the robot will move. When it has traveled this distance, a set of Move blocks puts the robot in a good position to perform another scan using the Search My Block. Then the robot will turn around and grab and lift the object. After this, two parallel sequences are started: one to avoid walls, the other to end the program whenever a loud sound is made.

Step 9: Create a new document. Save the program and name it *GrabBot*. Then place five blocks configured as shown in Figure 9-14, and connect the blocks with wires in the appropriate block plugs.

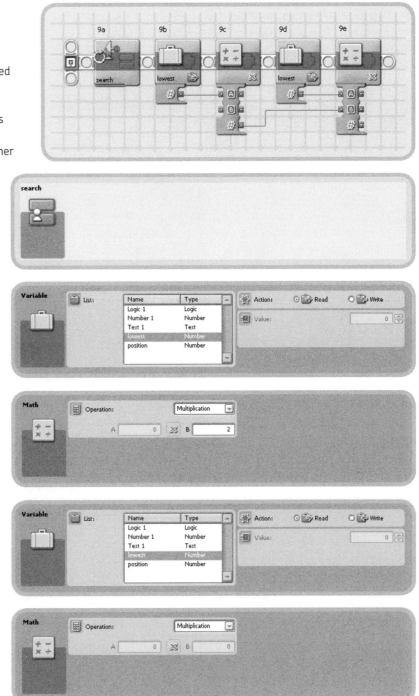

Figure 9-14: This is the configuration of the blocks placed in step 9.

Step 10: Place four blocks configured as shown in Figure 9-15. The output value of the Math block is wired into the Move block.

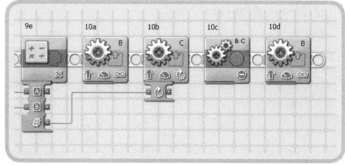

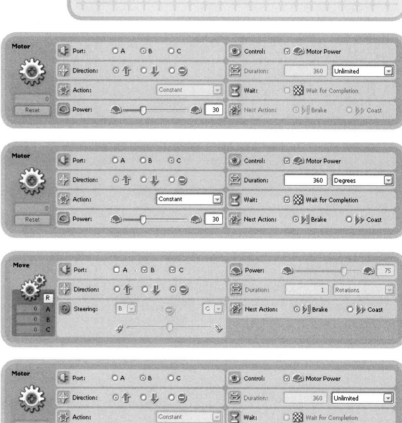

Figure 9-15: This is the configuration of the blocks placed in step 10.

Step 11: Place five blocks configured as shown in Figure 9-16.

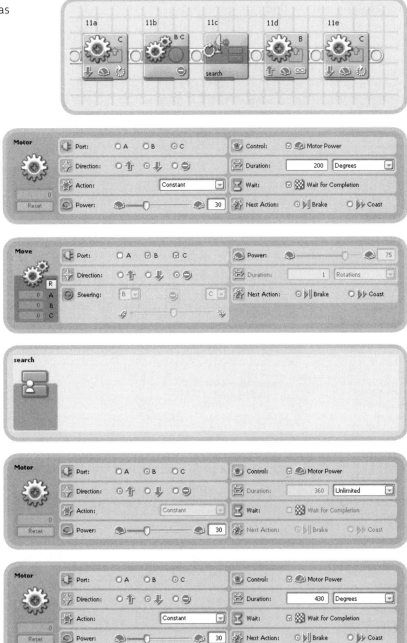

Figure 9-16: This is the configuration of the blocks placed in step 11.

NOTE Blocks 11d and 11e make the robot spin 180 degrees to point the grabber toward the object. On my wooden floor this is a 430-degree turn of the NXT motor, so my Motor block 11e is configured to turn 430 degrees. This value may vary for your floor because of a different amount of friction. The best way to find the right value is to run the finished program and modify the value until it works correctly.

Step 12: Place four blocks configured as shown in Figure 9-17.

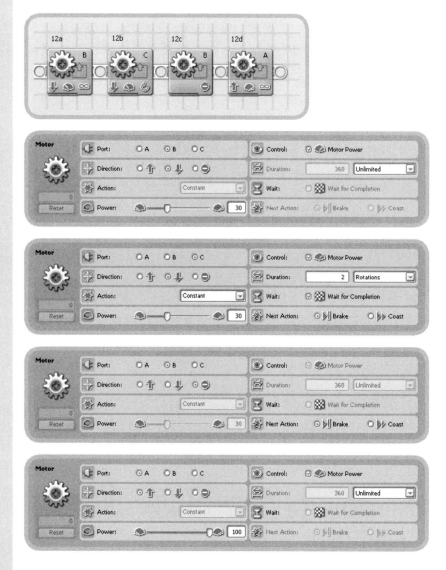

Figure 9-17: This is the configuration of the blocks placed in step 12.

Step 13: Place two blocks and a Loop block configured as shown in Figure 9-18.

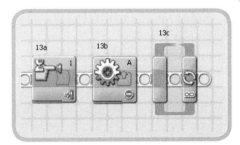

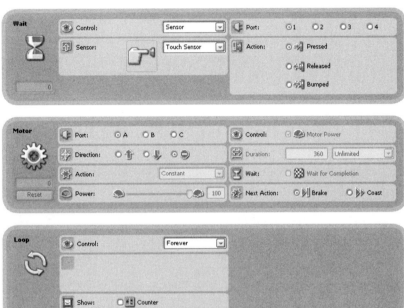

Figure 9-18: This is the configuration of the blocks placed in step 13.

Step 14: Place four blocks configured as shown in Figure 9-19.

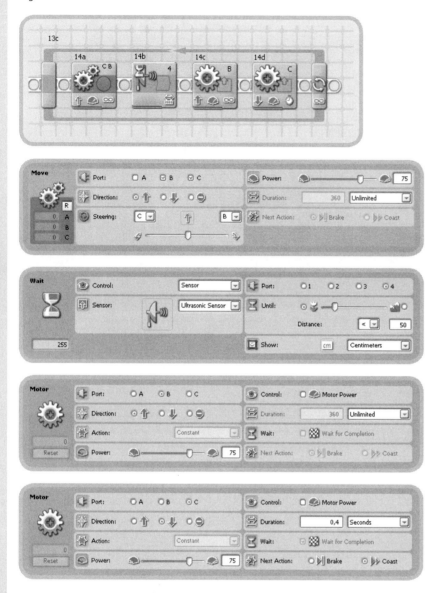

Figure 9-19: This is the configuration of the blocks placed in step 14.

Step 15: Place two blocks and a Loop block configured as shown in Figure 9-20. Then use the SHIFT key to drag a beam from 13b to 15a.

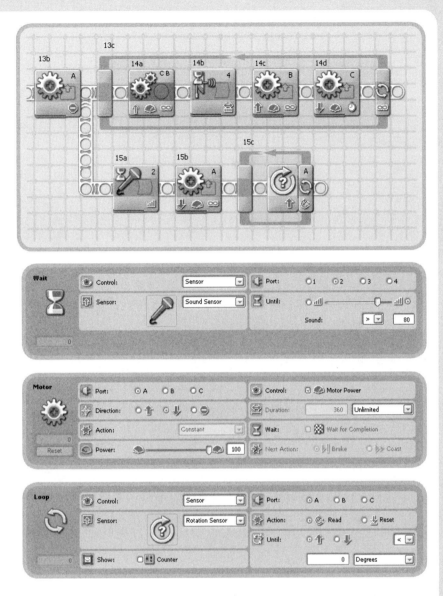

Figure 9-20: This is the configuration of the blocks placed in step 15.

Step 16: Place three blocks configured as shown in Figure 9-21.

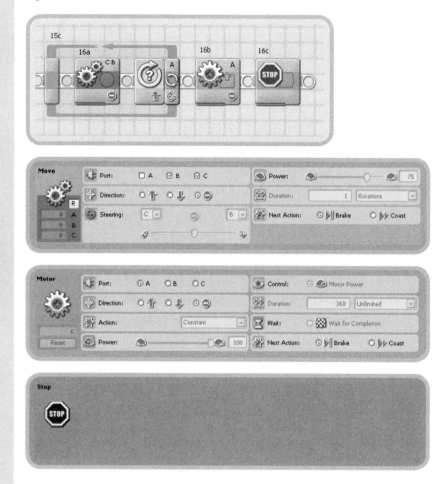

Figure 9-21: This is the configuration of the blocks placed in step 16.

Now save the program, compile, and run!

further exploration

Instead of using an Ultrasonic Sensor to find objects, you could use a Light Sensor to find dark corners or bright spots in a room. To program this behavior, you can use the same My Block with some simple modifications to make it search for the lowest light setting instead of the shortest distance.

First, save the Search My Block under a new name, such as *FindDarkCorner*. Then replace the Ultrasonic Sensor block (block 3b) with a Light Sensor block, and program the robot to grab an object and then look for a dark corner in which to place it.

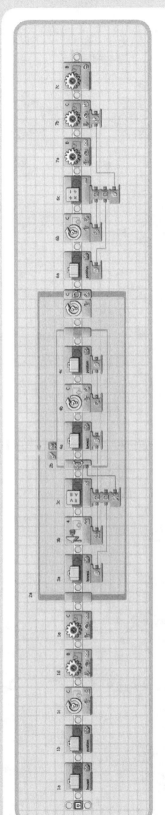

Figure 9-22: This is an overview of the Search My Block.

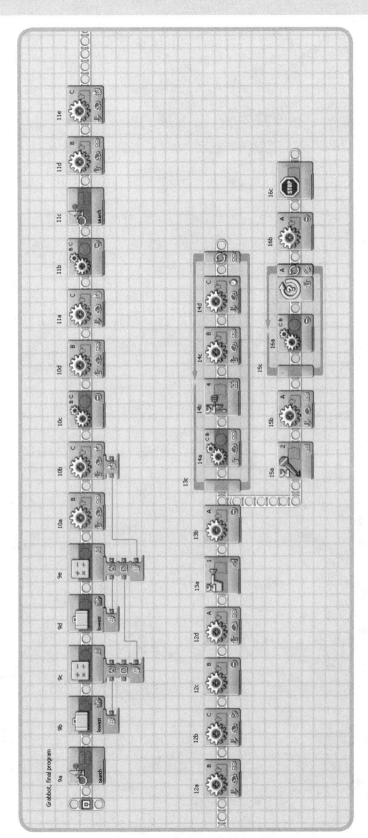

Figure 9-23: This is the completed program. (Note that the actual program isn't split into two halves.)

10

the bike: an NXT bicycle that can steer and move by itself

In this chapter you will learn how to build a bicycle (with training wheels) that moves forward and backward and steers to avoid obstacles.

To maintain symmetry and balance, we'll use two motors to power the rear wheel of the Bike. To improve steering and maneuverability, a third motor and a gear train are also positioned at the rear, reducing the weight on the front wheel. An array of worm gears, bevel gears, and axles transfer the motor's power from the rear to the front of the Bike. Two stabilizers act as training wheels to prevent the Bike from falling over.

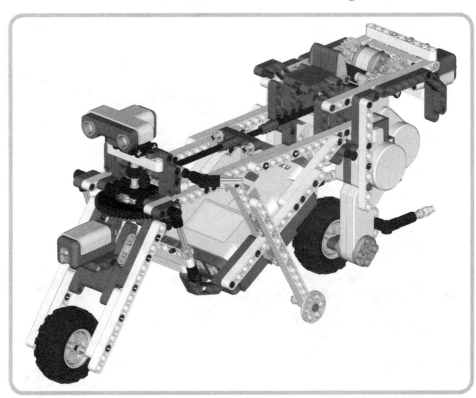

Figure 10-1: The Bike

building the bike

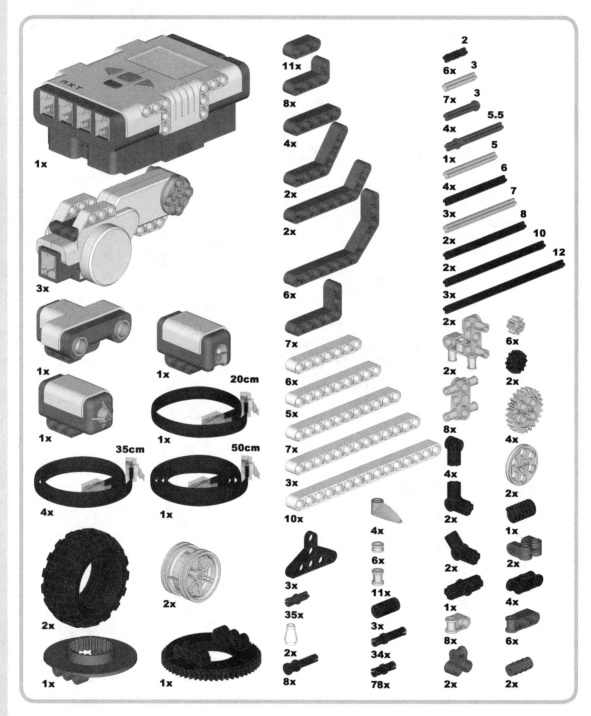

Figure 10-2: Bill of Materials

Let's start with the front of the Bike.

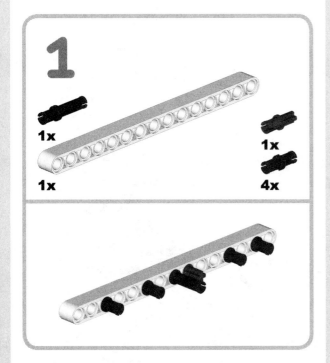

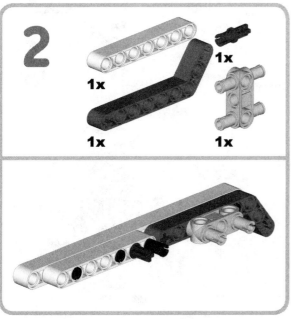

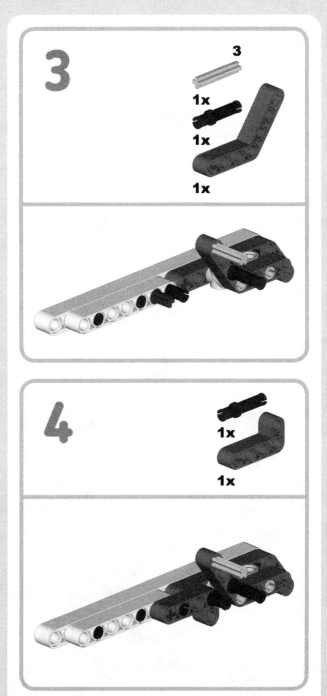

5

1x

1x

1x

6

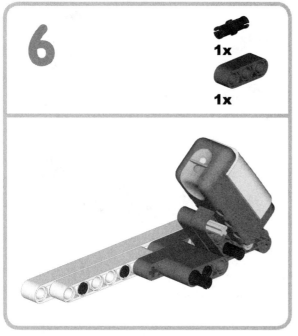

1x

1x

7

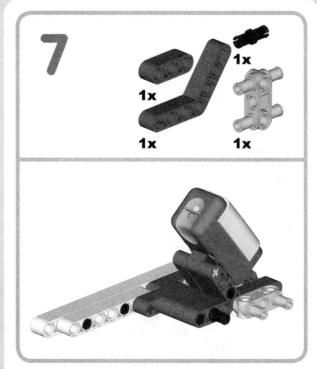

1x

1x

1x

1x

8

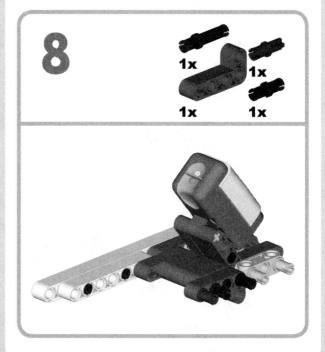

1x

1x

1x

1x

9

1x

2x

1x

10

2x

1x

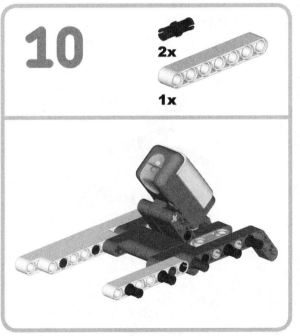

11

1x

12

2x   2x   2x

the handlebar and the front fork

1

5.5

1x   1x

1x

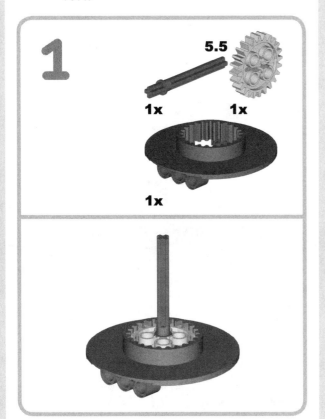

2

1x

3   1x

4   5

2x

5

4x

6

1x 1x 1x

7

2x

1x 2x 2x 2

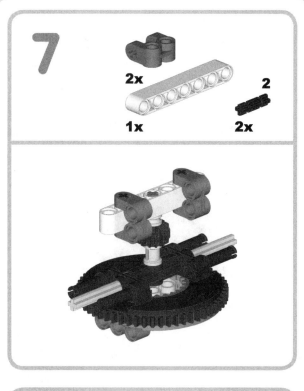

8

2x

2x

9

2x

2x

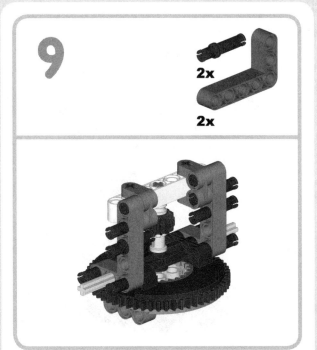

11

2x

2x

10

2x

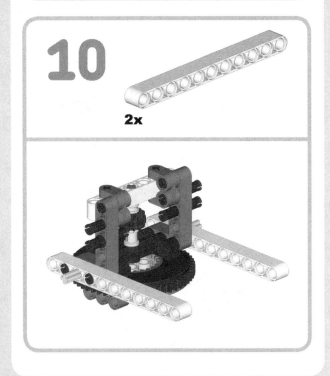

1x

1x

1x

12

12

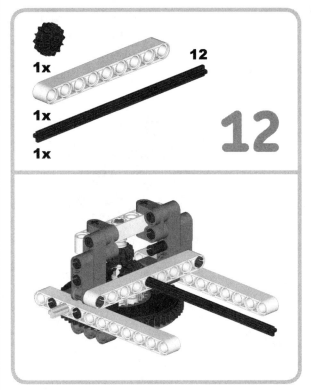

13

2x

2x

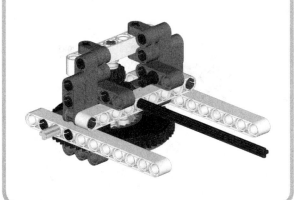

15

1x

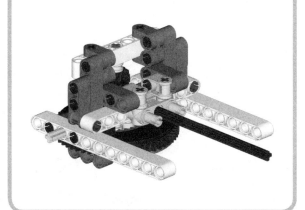

14

2x

2x

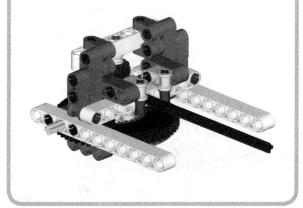

16

2x

2x

17

2x · 2 · 2x · 1x · 1x

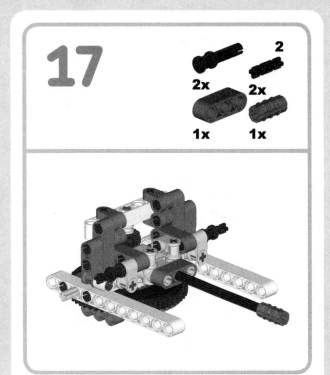

18

3 · 2x · 2x

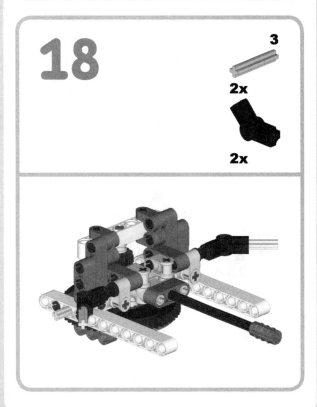

19

4x · 2x

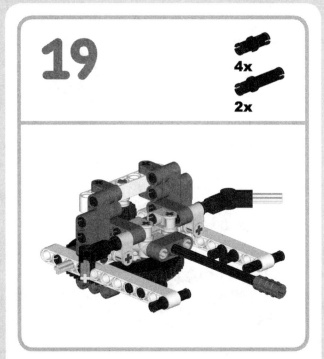

20

2x · 2x · 2x · 2x

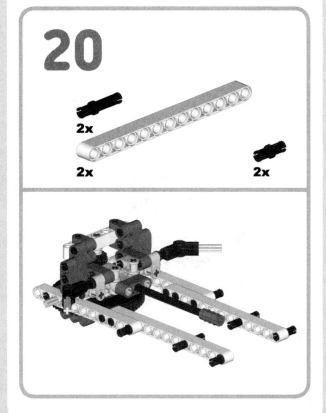

21

2x

2x

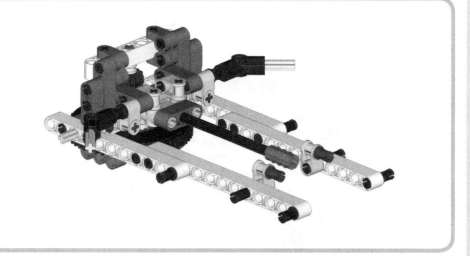

22

6

1x

1x

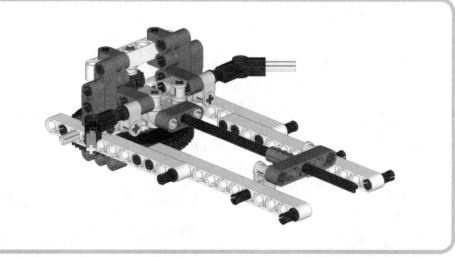

23

2x

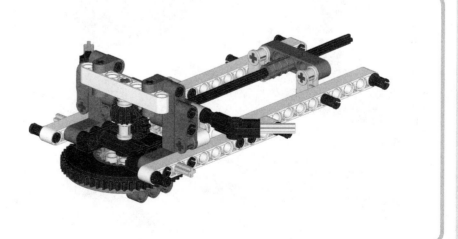

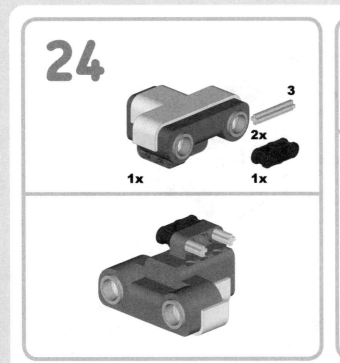

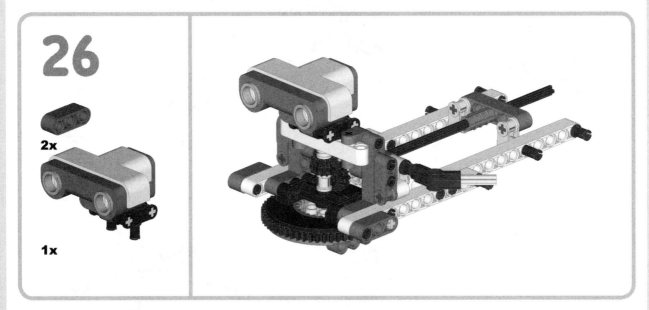

Join the handlebar and the front fork together, then add the front wheel as shown in the following steps.

27

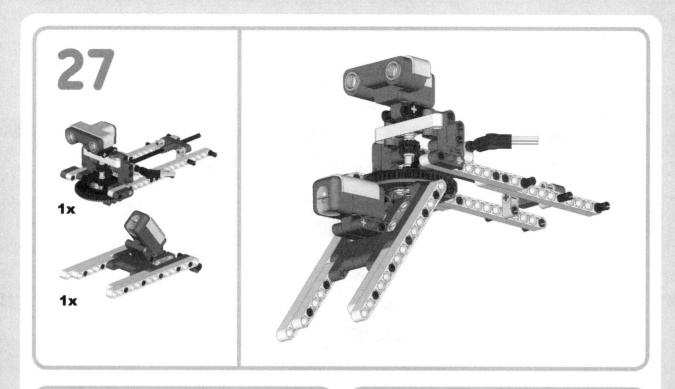

1x

1x

28

8

1x

1x

1x

29

1x

1x

1x

30

1x

the drive and the gear train

The drive and gear train will be positioned at the rear of the Bike to control the front-wheel steering.

1

2x   2x   10

1x   1x

2

6

1x

2x

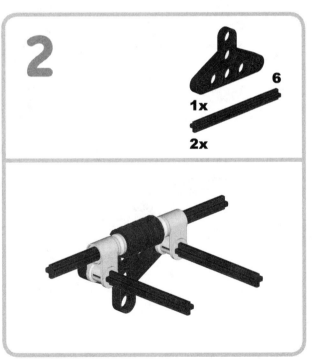

3

12

1x

1x

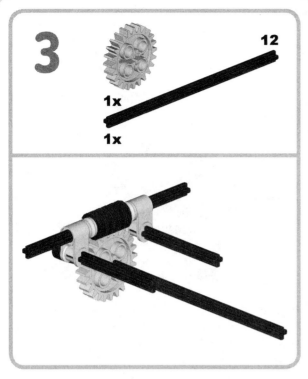

4

1x

2x

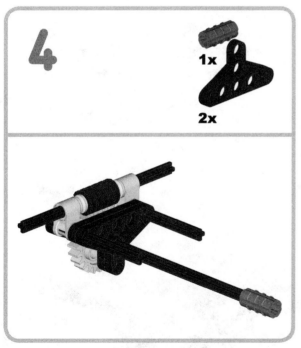

5

2x

6

1x   2x   2x

7

2x

8

2x 2x

2x

9

1x

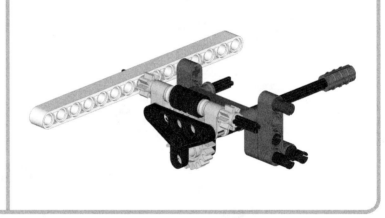

10

10

2x

1x 2x

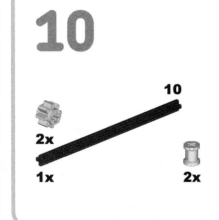

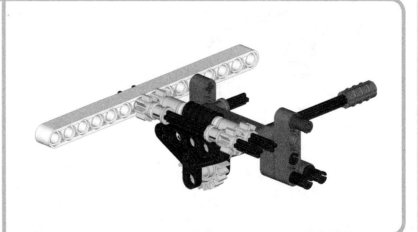

11

1x

1x

10

1x

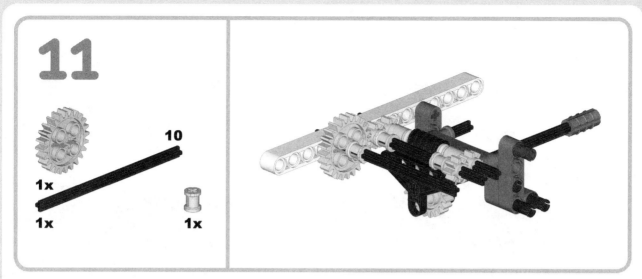

12

1x

1x

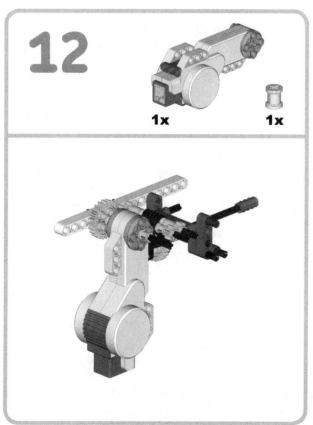

1x

1x

13

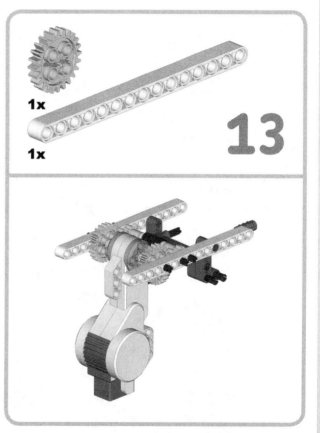

14

3
2x
1x

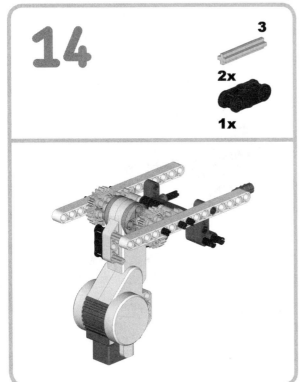

16

2x
1x

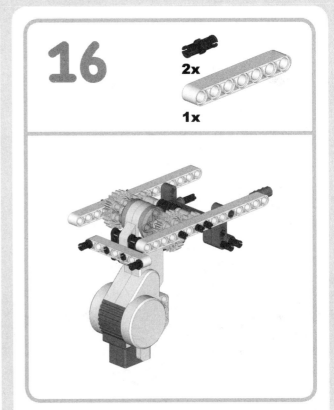

15

2x
1x

17

2x
1x   2x

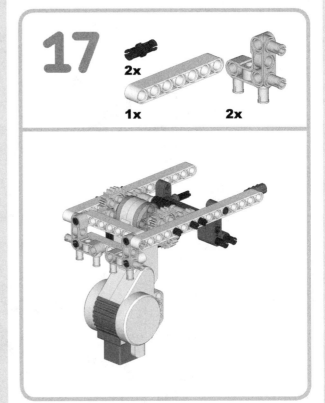

18

2x

2x

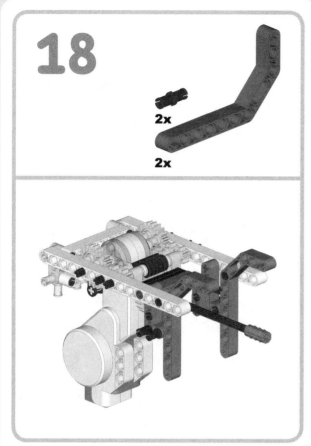

20

2x

2x

2x

2x

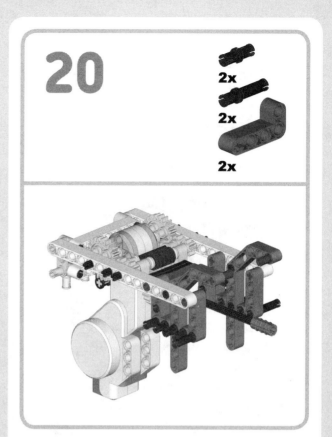

19

4x

2x

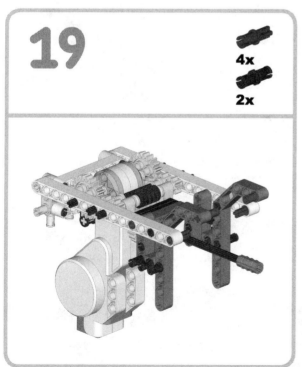

21

1x

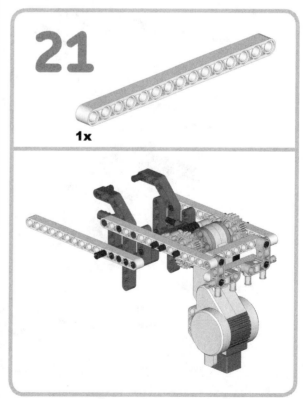

22

1x 2x 1x

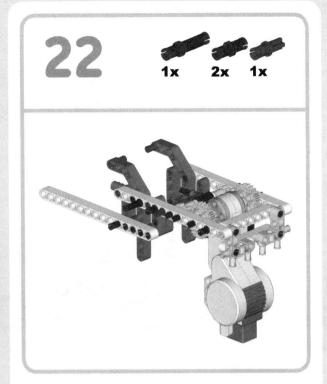

23

1x 1x 1x

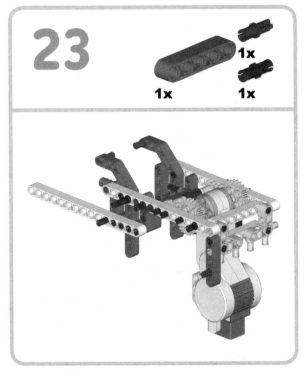

24

1x 1x 1x

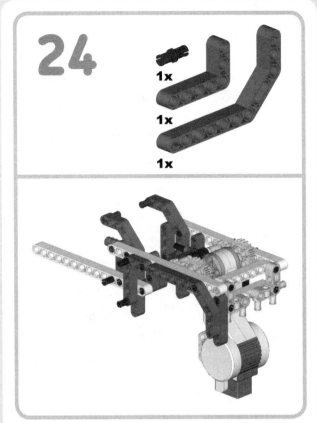

1x 1x 2x 1x 1x

25

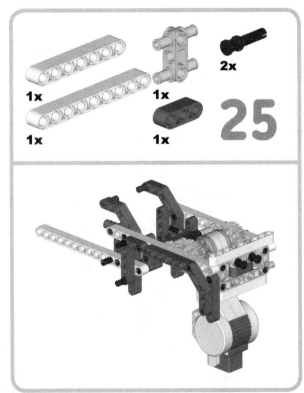

26

2x 3x 1x

1x

1x

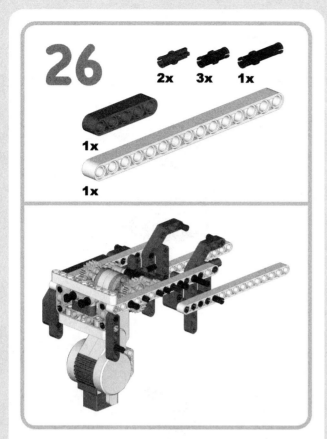

27

1x

1x

1x

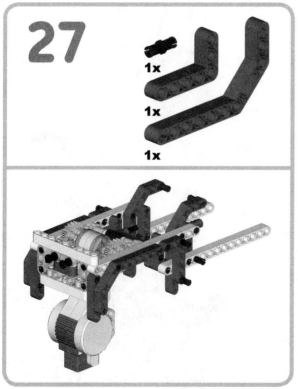

the seat

1

1x

1x

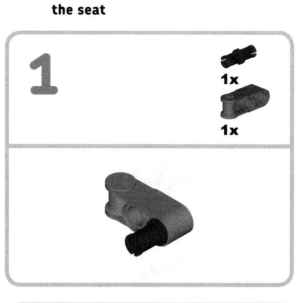

2

2x

1x

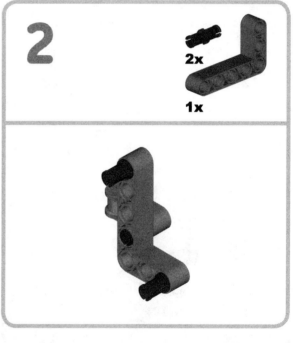

3

2x

1x

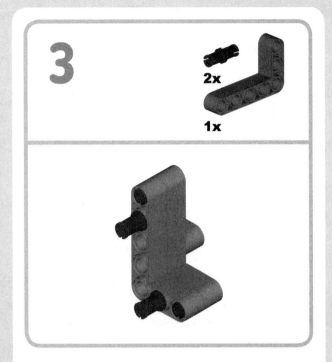

4

1x

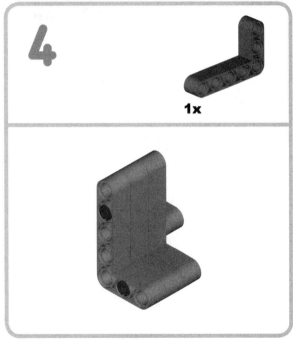

5

4x 1x

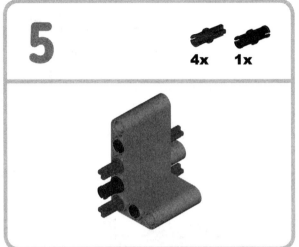

6

4x

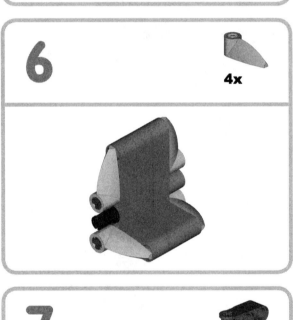

7

1x

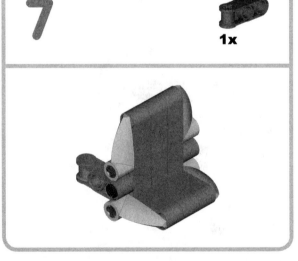

8

1x

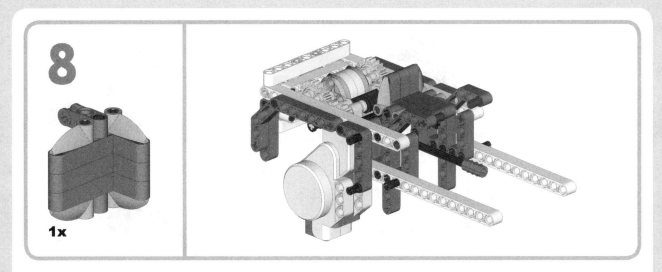

the wheel and drive motors

1

1x   2x

1x   1x

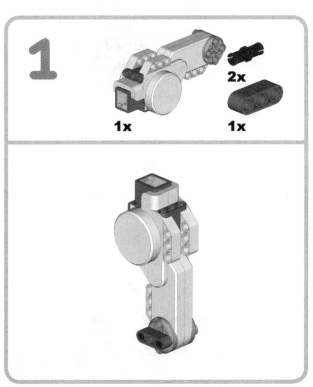

2

1x   8

1x   1x

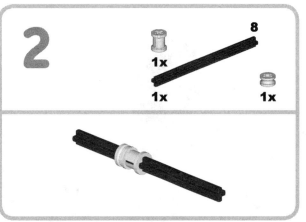

3

1x

1x   1x

4

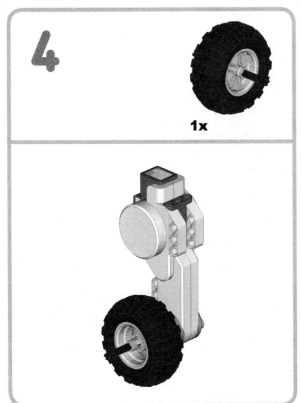

1x

5

2x

1x

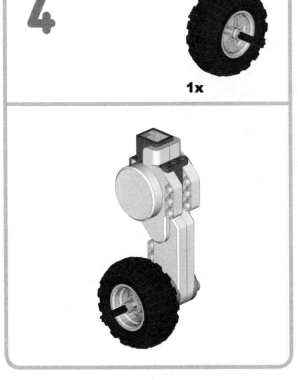

6

4x

1x

7

1x

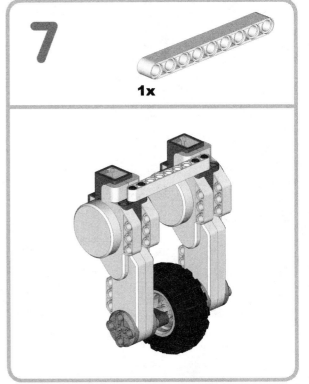

8

2x
2x

9

3
2x
2x

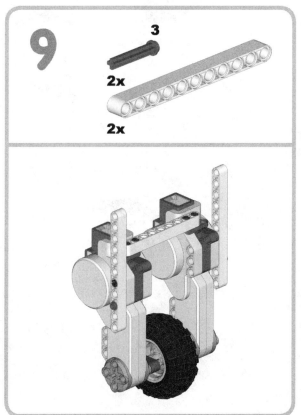

10

1x
5
1x

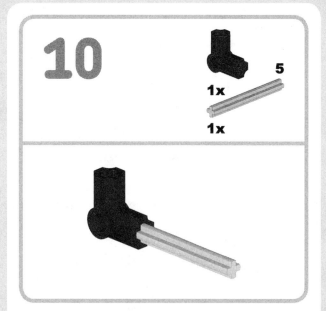

11

1x
1x
1x
1x

Build two of these.

12

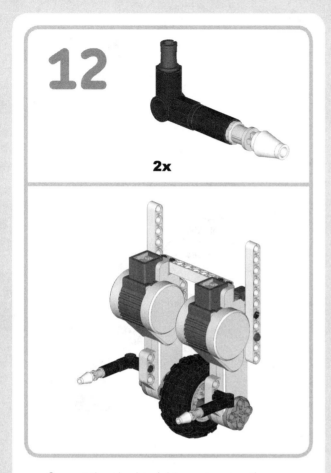

2x

Connect the wheel and drive motors together as shown in the following steps.

13

1x

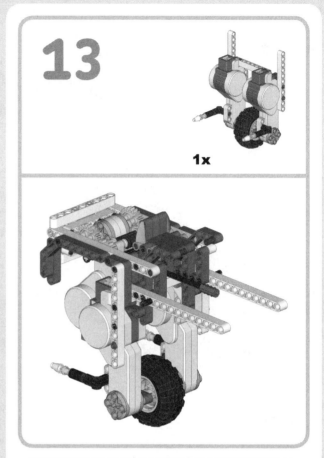

14

2x

the final assembly

First, add a Touch Sensor to the rear of the Bike.

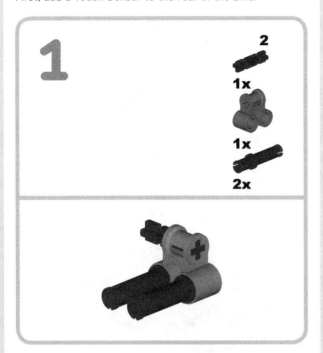

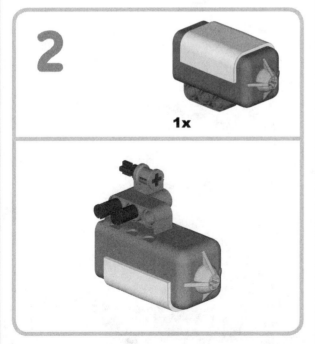

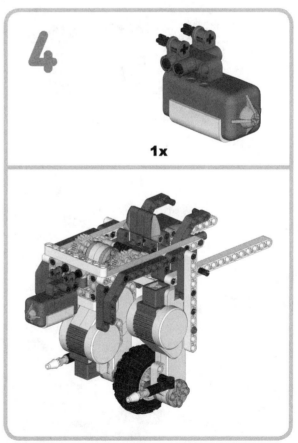

Next, construct the middle frame of the Bike and attach the NXT Brick.

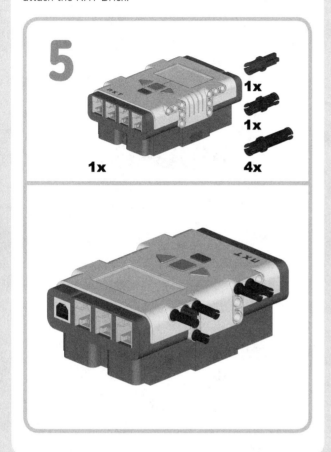

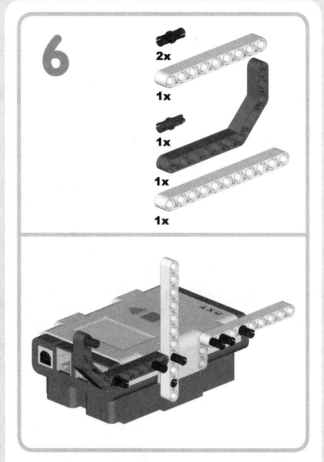

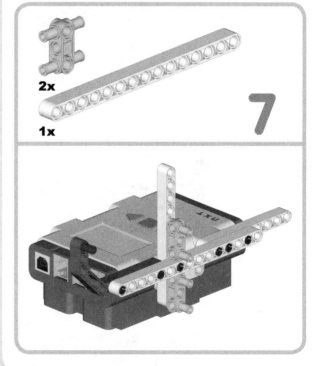

8

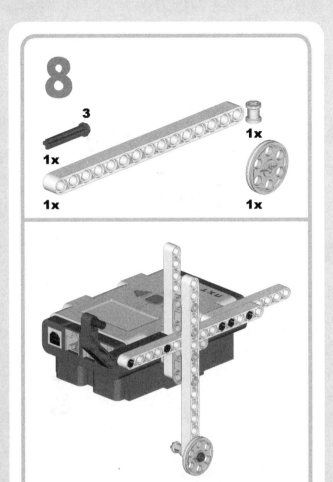

3
1x 1x
1x 1x

10

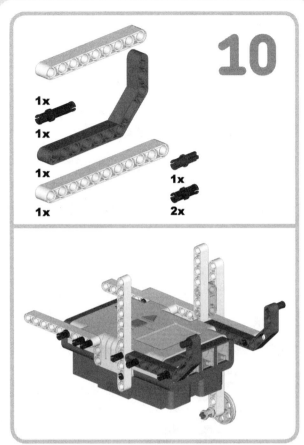

1x
1x
1x 1x
1x 2x

9

3x 1x 1x

2x

11

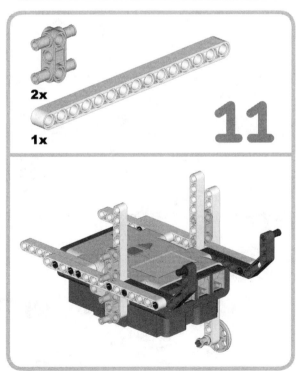

1x

12

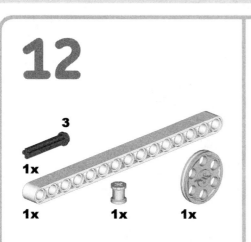

3
1x

1x   1x   1x

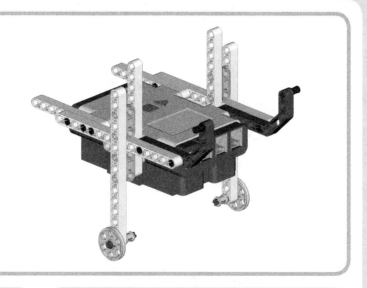

13

2x
1x

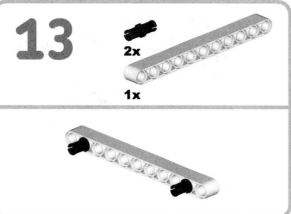

14

7
2x
1x   2x

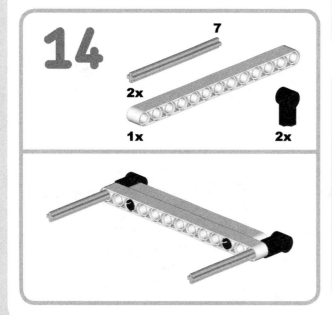

15

2x
2x

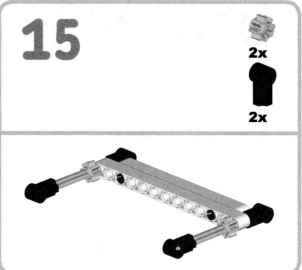

16

2x

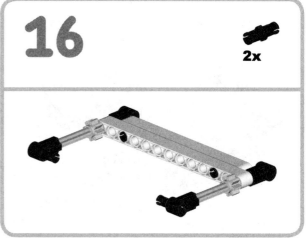

17

1x

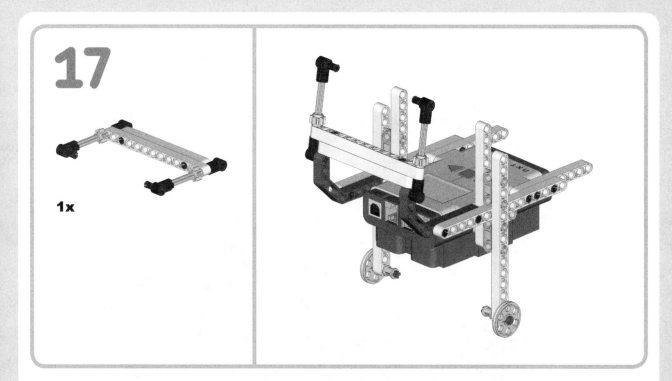

Finally, connect the front, middle, and rear sections as shown in the following steps.

18

1x

1x

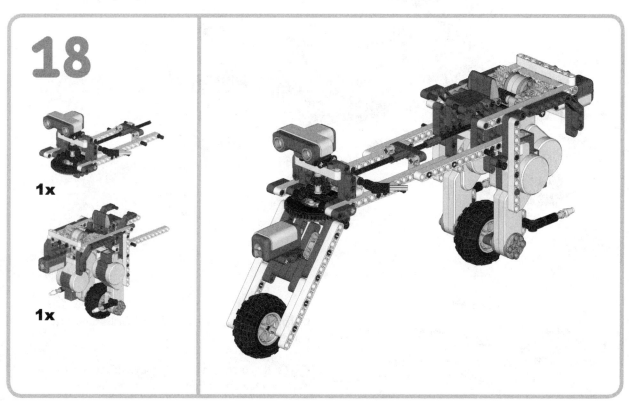

19

1x

connecting the cables

Connect the motor that steers the front wheels to output port A. Plug the motor on the left side of the rear wheel into output port B, and plug the motor on the right side of the rear wheel into output port C. The Touch Sensor should connect to input port 1, the Light Sensor should connect to input port 3, and the Ultrasonic Sensor should connect to input port 4.

NOTE If you deviate from these instructions, then you must change the ports in the program as well.

programming the bike

The following simple program will make your Bike run independently—it will ride around, turn randomly, and avoid objects. It's great fun to watch!

initialization

The first thing our program does is set initial values for the direction variable.

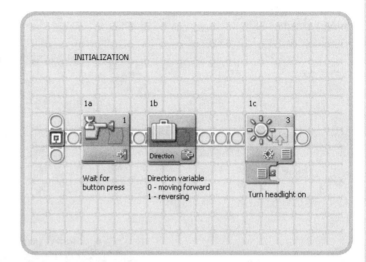

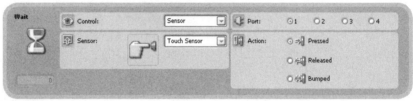

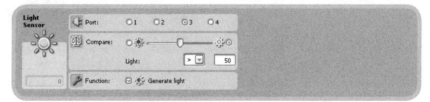

Figure 10-3: Configuration of blocks for initialization that await a Touch Sensor press and initialize variable

The program then begins executing the two parallel beams repeatedly.

obstacle avoidance and drive control

The first parallel beam directs the Ultrasonic Sensor to continually check for obstacles within 50 cm (20 inches) of the Bike's path. If an obstacle is detected, motors B and C reverse direction to avoid it.

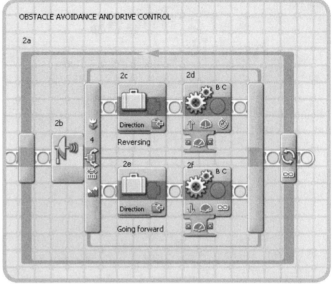

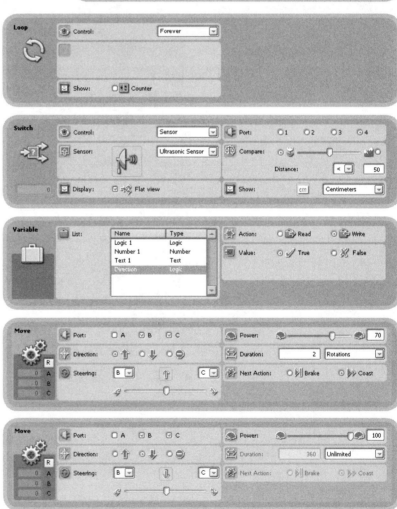

Figure 10-4: Configuration of blocks that control the motor drives and object detection

steering control

On the second beam, the front-wheel steering (controlled by motor A) is controlled using a random variable, turning the Bike left, right, and straight. While moving forward, the Bike changes direction every three seconds, maintains the new course for two seconds, then reverts back to a neutral position for one second. At the end of the sequence, it repeats the process, moving randomly until the battery runs out or something stops it.

A local variable stores the current direction the Bike is traveling. It is initially set to 0 (forward), and it remains 0 until it comes across an obstacle. If the current Bike direction (as indicated by the direction variable) is *reverse*, the program does nothing.

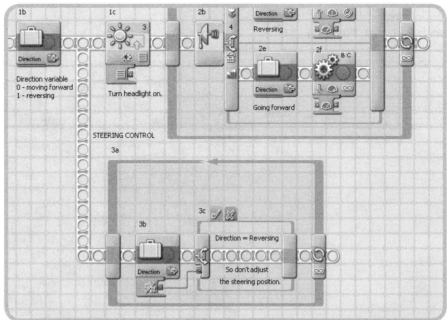

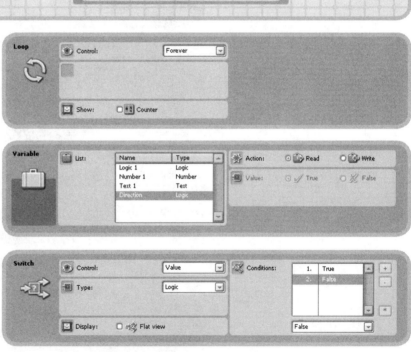

Figure 10-5: Configuration of blocks that control the random steering sequence

Figure 10-6: Switch block configuration for steering during reversing

However, if the current Bike direction is forward, then the following steering control beam is executed.

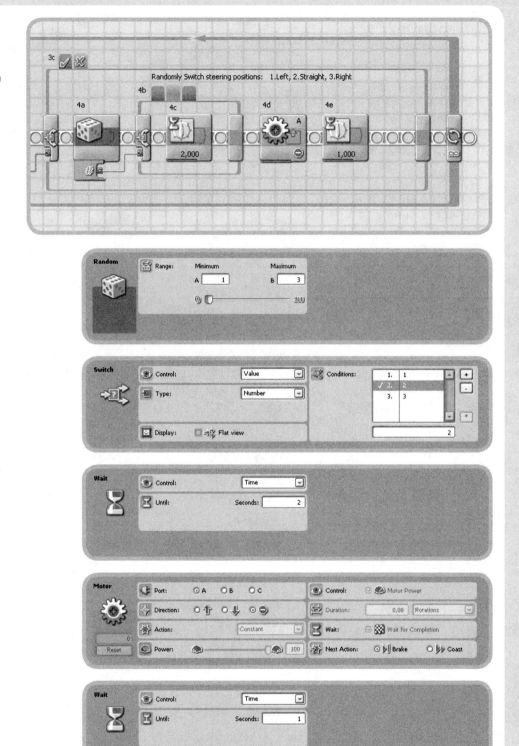

Figure 10-7: Switch block configuration for steering during forward motion

The configuration panel for the Switch block in Figure 10-7 switches the action depending on the random value generated by the Random block, as follows:

* If the random value is 2, the Bike continues moving straight ahead.
* If the random value is 1, the Bike turns left.
* If the random value is 3, the Bike turns right.

Straight

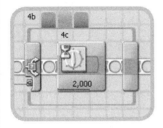

Figure 10-8: Configuration blocks for the sequence beam that moves the Bike forward in a straight line

Left

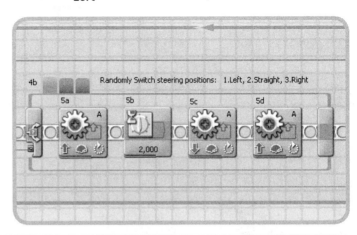

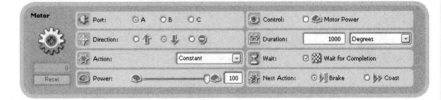

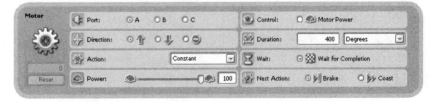

Figure 10-9: Configuration blocks for the sequence beam that turns the Bike left

Right

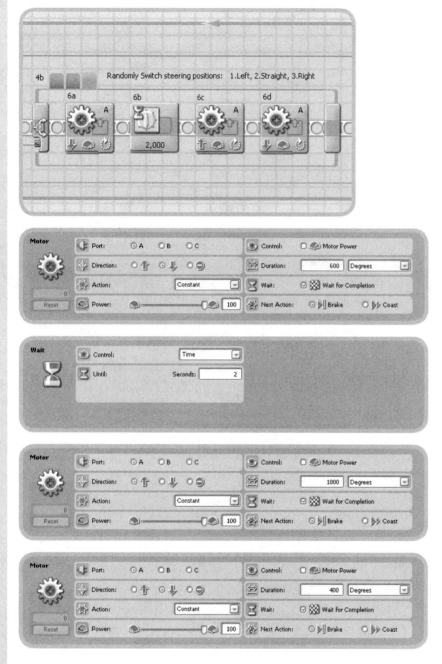

Figure 10-10: Configuration blocks for the sequence beam that turns the Bike right

When the random steering control and the obstacle avoidance processes are combined, it results in the Bike continually driving around in random directions and avoiding obstacles.

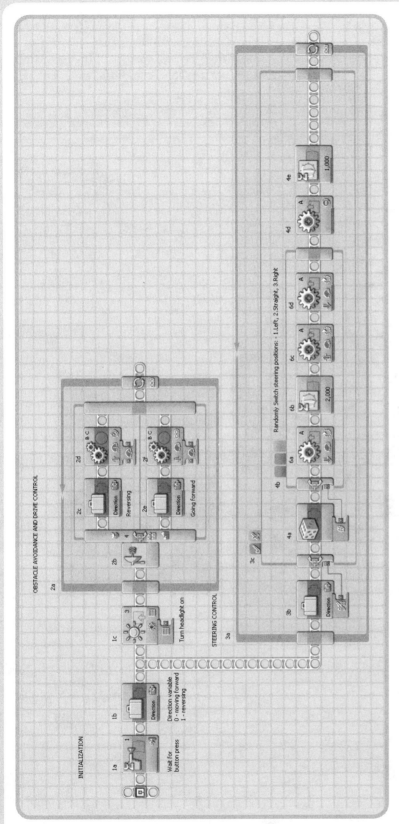

Figure 10-11: The complete Bike program

further exploration

For bonus material to help you modify the Bike, we have provided additional building instructions and programs that you can download from the Books section of The NXT STEP blog (*http://thenxtstep.com/book/downloads*).

This book is about creating models out of a single NXT kit. However, once you have built your Bike, there is nothing stopping you from modifying it using additional TECHNIC pieces and extra NXT sensors and motors. For example:

* Can you make the Bike self-balancing?
* Can you make the Bike follow a straight line?

Figure 10-11: An example of a standard Bike modified with TECHNIC bike wheels, a Light Sensor (for line following), and a Gyro Sensor (for balancing)

colophon

LEGO MINDSTORMS NXT One-Kit Wonders was laid out in Adobe InDesign. The font is Chevin.

The book was printed and bound at Malloy Incorporated in Ann Arbor, Michigan. The paper is Glatfelter Spring Forge 60# Smooth Eggshell, which is certified by the Sustainable Forestry Initiative (SFI). The book uses a RepKover binding, which allows it to lay flat when open.

The Unofficial LEGO® MINDSTORMS® NXT Inventor's Guide

by DAVID J. PERDUE

The Unofficial LEGO MINDSTORMS NXT Inventor's Guide will teach you how to successfully plan, construct, and program robots using the MINDSTORMS NXT set, the powerful robotics kit designed by LEGO. This book begins by introducing you to the NXT set and discussing each of its elements in detail. Once you are familiar with the beams, gears, sensors, and cables that make up the NXT set, the author offers practical advice that will help you plan, design, and build robust and entertaining robots. The book goes on to cover the NXT-G programming environment, providing code examples and programming insights along the way. Rounding out the book are step-by-step instructions for building, programming, and testing six complete robots that require only the parts in the NXT set; an NXT piece library; and an NXT-G glossary.

OCTOBER 2007, 320 PP., $29.95
ISBN 978-1-59327-154-1

The LEGO® MINDSTORMS® NXT Idea Book
Design, Invent, and Build

by MARTIJN BOOGAARTS, JONATHAN A. DAUDELIN, BRIAN L. DAVIS, JIM KELLY, DAVID LEVY, LOU MORRIS, FAY RHODES, RICK RHODES, MATTHIAS PAUL SCHOLZ, CHRISTOPHER R. SMITH, *and* ROB TOROK

With chapters on programming and design, CAD-style drawings, and an abundance of screenshots, *The LEGO MINDSTORMS NXT Idea Book* makes it easy for readers to master the LEGO MINDSTORMS NXT kit and build the eight example robots. Readers learn about the NXT parts (beams, axles, gears, and so on) and how to combine them to build and program working robots like a slot machine (complete with flashing lights and a lever), a black-and-white scanner, and a robot DJ. Chapters cover using the NXT programming language (NXT-G) as well as troubleshooting software, sensors, Bluetooth, and even how to create an NXT remote control. LEGO fans of all ages will find this book an ideal jumping-off point for doing more with the NXT kit.

SEPTEMBER 2007, 368 PP., $29.95
ISBN 978-1-59327-150-3

Forbidden LEGO®
Build the Models Your Parents Warned You Against!

by ULRIK PILEGAARD *and* MIKE DOOLEY

Written by a former master LEGO designer and a former LEGO project manager, this full-color book showcases projects that break the LEGO Group's rules for building with LEGO bricks—rules against building projects that fire projectiles, require cutting or gluing bricks, or use nonstandard parts. Many of these are back-room projects that LEGO's master designers build under the LEGO radar, just to have fun. Learn how to build a catapult that shoots M&Ms, a gun that fires LEGO beams, a continuous-fire ping-pong ball launcher, and more! Tips and tricks will give you ideas for inventing your own creative model designs.

AUGUST 2007, 192 PP. *full color*, $24.95
ISBN 978-1-59327-137-4

The Unofficial LEGO® Builder's Guide

by ALLAN BEDFORD

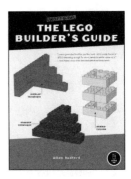

The Unofficial LEGO Builder's Guide combines techniques, principles, and reference information for building with LEGO bricks that go far beyond LEGO's official product instructions. Readers discover how to build everything from sturdy walls to a basic sphere, as well as projects including a mini space shuttle and a train station. The book also delves into advanced concepts such as scale and design. Includes essential terminology and the Brickopedia, a comprehensive guide to the different types of LEGO pieces.

SEPTEMBER 2005, 344 PP., $24.95
ISBN 978-1-59327-054-4

The LEGO® MINDSTORMS® NXT Zoo!
An Unofficial, Kid-Friendly Guide to Building Animals with LEGO MINDSTORMS NXT

by FAY RHODES

Whether you're just beginning with your LEGO MINDSTORMS NXT set or are already an expert, you'll have hours of fun with these animal-like models that walk, crawl, hop, and roll! The first part of the book introduces you to the NXT kit and reviews the parts you'll need in order to begin building. Next, you'll learn how to program with the NXT-G programming language, including how to make miniprograms called My Blocks that you can use to build larger programs. Finally, you'll learn how to build each robot and program it to act like its real animal cousins. Learn to build and program models like Ribbit, a jumping frog, Bunny, a hopping rabbit, and Sandy, a walking camel.

FEBRUARY 2008, 336 PP., $24.95
ISBN 978-1-59327-170-1

PHONE:
800.420.7240 OR
415.863.9900
MONDAY THROUGH FRIDAY,
9 AM TO 5 PM (PST)

FAX:
415.863.9950
24 HOURS A DAY,
7 DAYS A WEEK

EMAIL:
SALES@NOSTARCH.COM

WEB:
WWW.NOSTARCH.COM

MAIL:
NO STARCH PRESS
555 DE HARO ST, SUITE 250
SAN FRANCISCO, CA 94107
USA

updates

Visit *http://nostarch.com/nxtonekit.htm* for updates, errata, and other information.